The Crisis of Global Yo Unemployment

T0187791

Since the economic and financial crisis of 2008, the proportion of unemployed young people has exceeded any other group of unemployed adults. This phenomenon marks the emergence of a *laborscape*. This concept recognizes that, although youth unemployment is not consistent across the world, it is a coherent problem in the global political economy.

This book examines this crisis of youth unemployment, drawing on international case studies. It is organized around four key dimensions of the crisis: precarity, flexibility, migration, and policy responses. With contributions from leading experts in the field, the chapters offer a dynamic portrait of unemployment and how this is being challenged through new modes of resistance. This book provides cross-national comparisons, both ethnographic and quantitative, to explore the contours of this laborscape on the global, national, and local scales. Throughout these varied case studies is a common narrative from young workers, families, students, volunteers, and activists facing a new and growing problem.

This book will be an imperative resource for students and researchers looking at the sociology of globalization, global political economy, labor markets, and economic geography.

Tamar Mayer is the Robert R. Churchill Professor of Geosciences and Director of both the Rohatyn Center for Global Affairs and the International and Global Studies Program at Middlebury College, USA.

Sujata Moorti is the Charles A. Dana Professor of Gender, Sexuality, and Feminist Studies at Middlebury College, USA.

Jamie K. McCallum is an associate professor of Sociology at Middlebury College, USA.

Routledge Studies in Human Geography

This series provides a forum for innovative, vibrant, and critical debate within human geography. Titles will reflect the wealth of research which is taking place in this diverse and ever-expanding field. Contributions will be drawn from the main sub-disciplines and from innovative areas of work which have no particular sub-disciplinary allegiances.

Geographical Gerontology
Edited by Mark Skinner, Gavin Andrews, and Malcolm Cutchin

New Geographies of the Globalized World
Edited by Marcin Wojciech Solarz

Creative Placemaking
Research, Theory, and Practice
Edited by Cara Courage and Anita McKeown

Living with the Sea
Knowledge, Awareness, and Action
Edited by Mike Brown and Kimberley Peters

Time Geography in the Global Context
An Anthology
Edited by Kajsa Ellegård

Space, Grief, and Bereavement
Consolationscapes
Edited by Christoph Jedan, Avril Maddrell, and Eric Venbrux

The Crisis of Global Youth Unemployment
Edited by Tamar Mayer, Sujata Moorti, and Jamie K. McCallum

For more information about this series, please visit: www.routledge.com/Routledge-Studies-in-Human-Geography/book-series/SE0514

The Crisis of Global Youth Unemployment

Edited by Tamar Mayer, Sujata Moorti, and Jamie K. McCallum

Routledge
Taylor & Francis Group

LONDON AND NEW YORK

First published 2019
by Routledge
2 Park Square, Milton Park, Abingdon, Oxon OX14 4RN

and by Routledge
52 Vanderbilt Avenue, New York, NY 10017, USA

First issued in paperback 2020

Routledge is an imprint of the Taylor & Francis Group, an informa business

British Library Cataloguing-in-Publication Data
A catalogue record for this book is available from the British Library

Library of Congress Cataloging-in-Publication Data
A catalog record has been requested for this book

ISBN 13: 978-0-367-58576-1 (pbk)
ISBN 13: 978-0-8153-7108-3 (hbk)

Typeset in Times New Roman
by Integra Software Services Pvt. Ltd.

For Robert E. "Bob" Prasch III
1959–2015

Contents

Figures

Tables

Contributors

Marcellus Andrews earned a PhD in economics from Yale University. He currently teaches economics at Bucknell University. Andrews' primary research and teaching interests include complex adaptive systems, macroeconomics, economic inequality, as well as analytical and philosophical approaches to economic justice.

Catherine Bowman is a doctoral candidate at the University of Colorado. Her scholarly focus is on migrant labor policy, the sociology of work, and race. Bowman holds a master's degree in international development from the University of Pittsburgh, where she was funded by a Fulbright fellowship to carry out research on Chilean human trafficking policy. In addition to her scholarly interests, she has several years of experience providing direct services and conducting policy work on behalf of immigrant victims of forced labor.

Svea Closser is an associate professor of anthropology and Director of the Global Health Program at Middlebury College. Her research focuses on the interaction between global health policy and local health systems. Recent projects include a seven-country study of polio eradication and health systems, funded by the Bill and Melinda Gates Foundation, and research on ground-level health staff in Ethiopia, funded by the National Science Foundation. She is the author of *Chasing Polio in Pakistan* (Vanderbilt University Press, 2010), *Understanding and Applying Medical Anthropology* (Routledge, 2016), and *Foundations of Global Health* (Oxford University Press, 2018), as well as many research articles.

Crispen Chinguno is a senior lecturer of sociology at the Sol Plaatje University in Kimberley, South Africa, and a research associate at the Society, Work and Development Institute (SWOP) at the University of the Witwatersrand, Johannesburg, South Africa. His research interests broadly interrogate power and resistance with a special focus on work, trade unions, social movements, and inequality. His academic and research interests emerged from his engagement with trade unions in Zimbabwe where he worked at the National Railways of Zimbabwe (NRZ).

Heidi Gottfried is an associate professor of sociology at Wayne State University. She has published several books and articles on the themes of gender, precarity, and work. *Gender, Work and Economy: Unpacking the Global Economy*

(John Wiley & Sons, 2013) explores the relationship between gender and work in the global economy. In her new book, *The Reproductive Bargain: Deciphering the Enigma of Japanese Capitalism* (Brill, 2015), Gottfried develops a gendered institutional analysis of work and employment in Japan. With Stephen Edgell and Edward Granter, she edited *The SAGE Handbook of the Sociology of Work and Employment* (2015), a landmark collection of original contributions by leading specialists from around the world.

Carlo Inverardi-Ferri is a post-doctoral research fellow in the Department of Geography and Global Production Networks Centre at the National University of Singapore. He completed his PhD at the School of Geography and the Environment at the University of Oxford in 2016, where he also worked as a teaching assistant within the Oxford Department of International Development. His broader intellectual commitment is the investigation of contemporary capitalism. Situated at the intersection of environmental and economic geography, his research examines how social relations, informal norms, and cultural elements shape economic activities and different perceptions of value, labor, development, and the environment.

Ange Bergson Lendja Ngnemzué holds a PhD in the philosophy of sciences from the University of Paris I, as well as a PhD in political science from the University of Paris VIII. He is the author of books and articles on immigration. He is an assistant professor and consultant in social sciences in Cameroon (Protestant University of Central Africa and University of Dschang), the United States (Boston University and Middlebury College), Japan (Kyoto University), and France (University of Paris VIII).

Stanford T. Mahati, a Zeit-Stiftung Ebelin und Gerd Bucerius "Settling into Motion" program alumni, received his PhD from the University of Witwatersrand, South Africa. He specializes in research on child migration, displacement, transnational migrant families, and designing and evaluating interventions targeting vulnerable children and their households. His current research around the intersections and navigation of migration, childhood, masculinity, and femininity is funded by the Wellcome Trust and is part of the Migration and Health Project.

Tamar Mayer is the Robert R. Churchill Professor of Geosciences at Middlebury College where she is the Director of both the Rohatyn Center for Global Affairs and the International and Global Studies Program. She is a political geographer whose research interests lie in the interplay among nationalism, landscape, and memory as it pertains to stateless ethnic nations, especially in the Middle East and Western China. She is the editor or co-editor of four books that focus on different dimensions of international and global crises, the most recent of which is *The Politics of Fresh Water* (Routledge, 2017).

Jamie K. McCallum is an associate professor of sociology at Middlebury College. His book, *Global Unions, Local Power* (Cornell University Press, 2013), won the best book prize from the labor section of the American Sociological Association in 2014.

Ciro Milione is an assistant professor at the Law and Economics School of Córdoba University in Spain. He holds an LLB and a PhD in constitutional law, for which he wrote a dissertation on "The Influence of the European Court for Human Rights on Italian and Spanish Constitutional Case Law" about the Due Process Right. He is also Professor of Comparative Political Institutions (Spain–USA) for PRESCHO, a consortium of six American schools for study abroad in Spain, as well as a researcher at the Center of Andalusian Studies Foundation ("Fundación Centro de Estudios Andaluces") in Seville. His research areas are human rights, regionalism, welfare states, and bioethics.

Sujata Moorti is the Charles A. Dana Professor of Gender, Sexuality, and Feminist Studies at Middlebury College. She specializes in analyzing media representations of race and gender in the United States. She has also published extensively on the transnationally circulating media of the South Asian diaspora and on Indian cinema.

Diane Negra is the Professor of Film Studies and Screen Culture and Head of Film Studies at University College Dublin. She is the author, editor, or co-editor of eight books, the most recent of which is *Gendering the Recession: Media and Culture in an Age of Austerity* (edited with Yvonne Tasker, Duke University Press, 2014).

Eleanor O'Leary is an assistant lecturer in Media and Communications at the Institute of Technology, Carlow. She is the author of *Youth and Popular Culture in 1950s Ireland* (Bloomsbury, 2018). Another recent publication is "Emigration, Return Migration and Surprise Homecomings" (Irish Studies Review, 2016), co-written with Professor Diane Negra (University College Dublin). O'Leary is currently completing a research project in conjunction with Negra and Dr. Anthony McIntyre (University College Dublin) funded by the Broadcasting Authority of Ireland on the topic of broadcasting Irish emigration in an era of global mobility.

Mariano D. Perelman has a PhD in social anthropology from the Universidad de Buenos Aires. He is a professor in the Doctoral Program in Social Sciences (UBA) and the Department of Anthropology of the Universidad de Buenos Aires. He is Adjunct Researcher at the Consejo Nacional de Investigaciones Científicas y Técnicas (CONICET, Argentina) and a researcher at the Urban Studies Area of the Instituto Gino Germini (IIGG-UBA).

Robert Prasch was Professor of Economics at Middlebury College where he taught courses on monetary theory and policy, macroeconomics, economic history, and the history of economic thought. He was the author of over 120 academic articles, book chapters, and reviews, along with numerous editorials and interviews in newspapers, radio, and online media, including *The Huffington Post, New Economic Perspectives, Translation Exercises, Salon*, and *Common Dreams*. His last book was *How Markets Work: Supply, Demand and the "Real World"* (Edward Elgar, 2008).

Acknowledgements

In 2014, we organized a conference on youth unemployment at Middlebury College. Drawing scholars from across the globe and a variety of disciplinary perspectives, we wished to outline how the economic meltdown since 2008 had manifested itself in different parts of the world and its particular impact on young people. This book is an outcome of the conversations that dominated the conference; we supplemented some of the presentations with specially solicited essays to ensure we could cover the social, political, and economic ramifications of the crisis. In the transition from conference to book manuscript, we lost unexpectedly two of our most important interlocutors on campus: Juana Gamero de Coca from the Spanish and Portuguese Department and Robert Prasch from the Economics Department, whose critique of unemployment policies and neoliberalism in general were pivotal to our arguments.

We thank Middlebury College's generosity and the many people who helped make the conference a success, specifically the Rohatyn Center for Global Affairs and its staff. We benefitted from the work of numerous research assistants: Meriely Amaral, Bree Baccaglini, Maria Bobbitt-Chertock, Laura Dillon, Sabina Latifovic, Marykate Melanson, Adriana Ortiz-Burnham, Elizabeth Sawyer, Greg Treiman, and Hyeon-Seok "Tom" Yu. Mary Bagg, thank you for your copyediting skills. Thanks as well to our editorial assistants at Routledge, Ruth Anderson and Pris Corbett.

We dedicate this book to Bob Prasch. We miss his wit, his belly-deep laugh, and his insights.

Tamar Mayer, Sujata Moorti, and Jamie K. McCallum

1 Global laborscapes of youth unemployment

Introduction

Tamar Mayer, Sujata Moorti, and Jamie K. McCallum

"You're fired!" During the early 2000s this was an oft-repeated phrase around the world as a long-term economic recession and high unemployment rates began to define the new millennium. It was also the catchphrase from the hit reality television show *The Apprentice* (on NBC from 2004 to 2015), which crassly used the prevalent fear of downsizing to draw audiences. The weekly ritual of seeing an apprentice being fired in an executive boardroom offered a glamorized image of scenes that everyday workers were encountering in their own lives. The show aired successfully in a number of global markets, and its adaptations were hits in at least 15 countries.[1] Its success can be attributed to its ability to capture an emergent structure of feeling, as Raymond Williams (1977) frames it. *The Apprentice* franchise and other pop-culture products tracked the changes in knowledge, expertise, and values characteristic of 21st-century post-industrial economies.

The Apprentice, as numerous cultural critics have noted, was the crucible that transformed Donald J. Trump from "a disgraced huckster who had trashed Atlantic City; a tabloid pariah to whom no bank would lend," into someone electable as president of the United States (Nussbaum 2017). While the show may have offered an image makeover for Trump, it and other reality TV shows of its ilk unmasked the logic of the prevalent vulture capitalism and the attendant anxieties about employment. The show's conceit was that, notwithstanding the persistent patterns of unemployment, meritocracy still defined Wall Street. Trump's ability to recast himself away from his role in failed business enterprises was a model lesson for how a country and its population could overcome the recession and fears of downsizing. More significantly, the show helped to restage key lessons on gender, race, and employment. Even as it gave voice to the precariousness and contingency of work patterns in neoliberal societies, the series foregrounded a social Darwinism: only the fittest survive. In broad brush strokes the series captured the key features of a risk society that the sociologist Ulrich Beck (1992) has theorized. The experience of unemployment itself is erased and the idea of being fired is presented as a "natural" outcome of competition.

The Apprentice tapped into a cultural zeitgeist. During the same period as the series' popularity, soaring unemployment rates were a characteristic feature in many parts of the world. Economists, policymakers, and cultural critics concurred that what distinguished the millennial economic crisis was the disproportionate representation of young adults in the category of the unemployed, sometimes by as much as a factor of three.

The International Labour Organization (ILO) estimates that over 65 million young people were unemployed in 2018,[2] an addition of about one million young people since 2014. Who belonged to this category of unemployed and what unemployment meant, however, varied from region to region and from country to country; as did the understanding of what age group defined the "youth" category. This definitional variability is one of the ideas foregrounded in *The Crisis of Global Youth Unemployment.* Together, the chapters chart the dominant and shared features of unemployment experienced around the world. Separately, each one points to particular ways in which individuals and countries have experienced the economic crisis.

The chapters in this anthology do not focus only on the economic dimensions of youth unemployment. Instead, they map out how persistent patterns of joblessness and precarity have reshaped cultures and individual subjectivities, and the ways in which people inhabit space. Policy responses to the economic crisis are also a key aspect of this cartography. As we write this introduction, economies around the world appear to have stabilized, but the number of unemployed remains striking. When we add to official unemployed statistics the numbers to account for young people who are not in school and are not actively in search of work, this figure (for 2017) is a staggering 676.6 million. The chapters collected in *The Crisis of Global Youth Unemployment* home in on this phenomenon in order to conceptualize its social, cultural, and political consequences. Through giving such simultaneous attention to the global and the local, and to the economic and cultural policies and social formations, *The Crisis in Youth Global Unemployment* examines the laborscapes of the new millennium.

Global laborscapes

To capture the radically disparate effects of globalization, the anthropologist Arjun Appadurai (1996) coined five terms: ethnoscapes, mediascapes, ideoscapes, finanscapes, and technoscapes. Building on this formulation, our use of the term "laborscape" captures the variegated effects of youth unemployment, including the new cultural and political formations that have come into being as a result. Intrinsically, laborscape recognizes that youth unemployment is not experienced similarly across the world; nor is the response to it similar. Rather, the term is designed to appreciate the heterogeneity of experiences associated with youth unemployment and the resultant sense of precarity that seems to have become a relentless feature of global society. Although the contributors to this volume do not use the term "laborscape," we understand their contributions provide the contours necessary for theorizing this concept.

For Appadurai, the suffix *-scape* suggests a vantage point from which to view the world and make connections between what seem like discrete phenomena. In a book like ours, it might be tempting to see so many varied perspectives from across the globe as no more than isolated "takes" on the problem. But a laborscape of global youth unemployment allows us to account for the disjuncture among the political, economic, geographic, and cultural dimensions of the crisis, forging a coherent theory out of different contexts. The popularity and longevity of *The Apprentice* is but one dimension of such a laborscape. In what follows we map the contours of this laborscape and how this collection addresses it: who is included and who is not, central categories and definitions, key stakeholders and actors, and the cultural and political ramifications of youth unemployment.

Contours of global unemployment

In *Capital*, Karl Marx (1976) posits that capitalism demands more youthful workers and fewer adult workers. The contemporary era of neoliberal capitalism seems to belie this assessment. The number of jobs available to young people has shrunk at an alarming rate since 2007. While young people constitute about 17% of the world's population, they represent 40% of its unemployed, including those who are looking for work, those who are not in school, and those who are not actively in search of work. Reports issued by the United Nations Office of the Secretary-General's Envoy on Youth echo these findings that youth unemployment is still at 13.1%, and this rate continued to be three times the adult unemployment rate in 2015.[3] The reason for the increase in global unemployment rates varies by place and over time, but the common thread appears to be neoliberal policies, privatization, and shrinking employment opportunities as well as an increase in young populations, especially in the non-Western world. Young people (aged 15–24) now make up close to one-fifth of the world's population and are the overwhelming majority (over 90%) in the developing world (Idris 2016: 2). With close to one billion young people expected to enter the labor market in the next decade, around 600 million new jobs will have to be created just "to keep unemployment rates constant" (Idris 2016: 2).

There is great variability within these statistics, however. For example, demographers characterize India as experiencing a youth bulge with nearly two-thirds of its population below the age of 35, but at present the jobless rate for male young people is at 10.3%; while in Greece, Spain, and the Palestinian Territories it is a vertiginously high 42.8%, 44.0%, and 41.7%, respectively.[4] In the United States, the unemployment rate among people aged 16–24 is at 9.5%, which is more than two and a half times the unemployment rate for those of all ages. In the European Union (EU), the youth unemployment rate ranges from 6.4% in Germany to the aforementioned 42.8% in Greece; and the overall rates of unemployment among young people in the EU are more than twice as high as they are among older generations.

In the Middle East and North Africa (MENA) region, where 60% of the Arab population is under the age of 30, and over one-third of them—more than 100 million—are aged 15–29 (ASDA'A Burson-Marsteller 2017), about one-quarter of the economically active young people are unemployed (25% in the Arab states and 30% in North Africa, in 2015). These rates are projected to increase in all MENA countries in the near future. In Africa as a whole, and particularly in southern parts of the continent, youth unemployment currently stands at above 30%, with South Africa at 57.4%, for example. But these rates do not tell the whole story, since many young people in Africa are underemployed and find their livelihood in the informal economy. Youth unemployment rates in Africa are particularly high in urban areas and affect mostly young people with at least secondary education (Baah-Boateng 2016).

These raw numbers are alarming but are complicated by other factors, among them: educational attainment, gender, race/ethnicity, ability, and religion. For instance, although gender differentials in youth unemployment rates are small at the global level, some regional differences are striking. In the advanced economies and East Asia, unemployment rates are lower for young women than for young men. But in the Middle East, South Asia, North Africa, southern Africa, and to a lesser extent Latin America and the Caribbean, the reverse is true. This is the case for all young women, but, paradoxically, the situation is more severe for young women with higher levels of education. In Turkey, for example, members of this demographic are three times more likely to be unemployed, and eight times more in Saudi Arabia. In South Asia, on the other hand, educated young women tend to be relegated to "volunteer" work while paid labor is reserved for men (see Chapter 4 in this volume).

Similarly, race and ethnicity play a significant role in youth employment prospects. In the United States, the jobless rate among African American and Hispanic young people is much higher than it is among white members of this age cohort. This is illustrated best by the solidity of the school-to-prison pipeline, which has produced the social landscape that Michelle Alexander (2012) terms the New Jim Crow. In Europe and Africa, discrimination against national minorities and migrants has increased exponentially. These conditions in turn have revitalized nascent nationalism sentiments,[5] which have radically destabilized existing global arrangements such as the EU and the African Union. While there are significant variations in how youth unemployment manifests itself across the world, all the resultant laborscapes signal a dramatic shift that needs to be understood and theorized as a coherent problem.

The statistics we have enumerated are worrisome on multiple counts. In almost every instance, the state is retrenching under a regime of economic austerity, and young people are increasingly forced to depend on families or a woefully inadequate patchwork of private charities. Existing patterns of disenfranchisement based on gender, class, and ethnicity are exacerbated in countries experiencing high levels of youth unemployment. Scholars posit provocative links between joblessness and increases in crime, political unrest, mental health problems, violence, and social exclusion. In addition, persistent youth unemployment is a harbinger of new political formations.

High rates of youth unemployment are more than just an economic concern; the ensuing poverty poses major social, political, and economic challenges, particularly to countries whose economies remain relatively weak. The combination of youth bulges and poor economic performance is particularly worrisome in countries where youth unemployment is high or rapidly rising. In such countries, this combination has led to instability, violence, social unrest, and a threat to national security (Nordås and Davenport 2013; Osakwe 2013). The Arab Spring that began in 2011 in Tunisia and spread to the rest of the Arab world, which resulted in political changes, social unrest, and even civil war, is a case in point.

In other words, high youth unemployment is a matter of national and global concern and requires immediate attention. The magnitude of this problem demands that we answer a critical question: How does the phenomenon of mass global youth unemployment impact the world today? From this general question, many others emerge:

- Are there differences in the experiences of unemployed young people in different countries?
- What are the effects of long-term youth unemployment on national culture, national identity, economic growth, and global labor flows?
- What is the relationship, if any, between educational attainment and youth unemployment?
- In what ways does the challenge of youth unemployment shape the experiences and rhythm of everyday life?
- What are the policy responses that address the concerns of long-term unemployed young people, especially their precarious conditions?

The authors of *The Crisis of Global Youth Unemployment*, writing from global and interdisciplinary perspectives, have made significant contributions to the cartography of this evolving laborscape.

This volume on youth unemployment addresses four central themes in the laborscape: precarity, flexibility, migration, and policy. Each theme provides a window through which to view a different aspect of this crisis. Hardly any chapter, however, is contained neatly within one theme. Rather, the intersecting natures of these contributions mirror the complexity of the problem.

Precarity

Social scientists in the last decade have undertaken a critical study of precarity, yet the quantitative explosion of this research threatens to broaden the concept beyond usefulness. The word itself has become shorthand for what seems increasingly like a terminal state in which social lives are unraveled by a breakdown in the welfare state, restricting access to good jobs and wages together with other necessities for the survival of individuals and social groups. In this book we use the term "precarity" most often in a way that is loyal to the original theorists who conceptualized an emerging condition for workers worldwide. As a result, for example, Guy Standing (2014) announced the arrival of the precariat: a political class of workers quite different from the classical conception of the white, male, industrial proletariat. Whether or not today's global youth workforce fits such a definition is less important than understanding how these young people's lives are structured by a precarious global capitalism, one that constantly undermines their gains and undercuts their prospects. The concept is also employed to discern a new cultural and political terrain for the growing numbers of those forced out of formal employment, yet who are still, in effect, working. In this sense, these young people are embedded in a precarious sociocultural context that informs not only their chances of finding work but also their expectations of what work means or signifies. Overall, precarity represents the empirical at the theoretical heart of today's global laborscape for young people.

Marcellus Andrews and Robert Prasch (Chapter 2) begin with an overview of the youth unemployment crisis, including a brief survey of what has been written previously and why that existing body of work is insufficient to understand the extent of the problem. In addition to providing some data to frame the discussions in several ensuing chapters, Andrews and Prasch argue persuasively that the set of solutions currently on offer, ranging from supply-side economics to austerity and "skills upgrading," is woefully inadequate. The chapter places the problem of youth unemployment in its proper historical context as the predictable outcome of years of stagnating wages, de-unionization, jobless growth, and demographic shifts.

For instance, the decades-long economic malaise in Japan has resulted in a series of new formations that we can now realize were precursors of the laborscape we chart in this book. Precarity, in the Japanese context, is both embodied and structural—affect and effect, as Heidi Gottfried contends in Chapter 3. Offering a gendered account of how economic transformations have altered workers' sense of citizenship, especially "company citizenship," Gottfried highlights how precarity has redefined the "hegemonic masculinity of a corporate-centered male-breadwinner system" (p. 38 in this volume). The quintessential figure of Japanese capitalism—the salaryman—was a generative myth that served to reinforce the gendered dichotomy of the male breadwinner and the female wife and mother. The economic crises of the 1990s brought into sharp relief the range of nonstandard employment that characterized the Japanese workforce; however, the masculine underpinnings of Japanese labor law and employment regulation remained intact. Chapter 3 highlights how today's Japanese male young people share the nonstandard employment practices that have characterized women's participation in the paid labor force. This in turn has radically revised social institutions that relied on the male breadwinner model and rendered even more fragile employment relations that were once hailed as the central pillar of Japanese capitalism.

Flexibility

In *The Order of Things*, Foucault slyly references Jorge Luis Borges's "Chinese encyclopedia" wherein:

animals are divided into: (a) belonging to the Emperor, (b) embalmed, (c) tame, (d) suckling pigs, (e) sirens, (f) fabulous, (g) stray dogs, (h) included in the present classification, (i) frenzied, (j) innumerable, (k) drawn with a very fine camelhair brush, (l) *et cetera*, (m) having just broken the water pitcher, (n) that from a long way off look like flies.

(Foucault 1970: xv)

The categorical crisis produced by such a list led Foucault into an investigation of knowledge formations. A similar conundrum of taxonomies characterizes the unemployment laborscape.

Work is defined contextually and historically, and most often in opposition to home, family, and leisure. As numerous sociologists have noted, work is a set of practices and behaviors that are determined through a specific space but cannot be captured within a single definition. This hermeneutical indeterminacy also shapes the forms of knowledge that exist about work, unemployment, and how policies are formed. Several chapters explore the ways in which the informal sector of labor—specifically waste picking and volunteer work—is invisible in official employment statistics. We have theorized this blurriness as a problem of *flexibility*, meaning the borders that typically shore up distinctions between work time and the rest of our lives are increasingly permeable. In post-industrial economies, flexible work schedules and contingent labor dominate. As a result of increasing neoliberalization, precarious work arrangements mean that young people often are forced to turn towards nontraditional avenues to make a living. Sometimes this means crisscrossing legal distinctions, worker classification issues, and cultural norms. Six chapters (3, 4, 5, 6, 7, and 10) explore the contours of those (flexible) boundaries. International volunteers, garbage pickers, students, street vendors, and migrant young people are among those caught in the liminal states that characterize the crisis for young workers today.

If categories of work and non-work are blurred, those distinguishing between unemployment and underemployment are even more subjective and tenuous. Statistics documenting the unemployed rely on criteria that represent only a fraction of the people in the labor force who are currently searching for work (Card 2011). The ILO defines the unemployed as those of working age who are currently: "a) without work [. . .]; b) currently available for work [. . .]; and c) seeking work."[6] The IMF and World Bank caution that "unemployment statistics can underestimate the true demands for jobs in an economy," but they do not offer any other definitional clarification (Oner 2017).

In Chapter 3, Gottfried underscores how the inclusion of nonstandard employment (temporary, part-time, and casual) further complicates our understanding of the laborscape. In the Japanese instance that she documents, these examples of contingent labor have contributed as much to the sense of precarity prevalent in Japan as has unemployment. Gottfried contends that the Japanese state is complicit in mystifying these categories of standard labor, precarious work, and unemployment. "Non-employment does not simply demarcate non-work or the absence of unemployment, but also glosses over non-market exchanges, ranging from reproductive labor to cooperative exchanges outside the orbit of traditional employment relations," she argues (p. 44 in this volume). In the United Kingdom, an analogous category of NEET (not in education, employment, or training) is used to address the fallout of neoliberal policies in post-industrial economies.

Similarly, there is no consensus among scholars about the best sources of information on unemployment statistics. In national contexts, government data are considered the most reliable, Europeans turn to data collected by the EU, and in international contexts the ILO is one of the most cited sources of global information. Each of these sources also defines the category of "youth" quite differently. While some agencies counted people between the ages of 18 and 25 in this category, others used the ages 18 to 34. In Chapters 4, 5, and 6, these categorical conundrums mark our statistics.

The gendered dimensions of work and unemployment classifications also shape the contours of the laborscape. Feminist scholars, such as Alice Kessler-Harris (1982) and Arlie Hochschild (1989), have highlighted how the arbitrary public-private divide not only shapes gender norms but also helps to reproduce the devaluation of certain forms of "feminized" labor. Several of the chapters in this anthology highlight the gendered nature of the unemployment laborscape. While men tend to be overrepresented in informal "public" work sites, such as waste collecting in China or Argentina, women tend to cluster in jobs that are classified as voluntary. These divisions are consequential; they shape public perceptions of work and the forms of disenfranchisement and alienation that individuals encounter. For instance, in many parts of the world, cultural ideas about femininity and domesticity ensure that the category of the unemployed does not include women who work for no pay. Many of the women Svea Closser (Chapter 4) encountered during her fieldwork in Pakistan worked from home and nearly none had full-time employment. Pakistani notions of propriety and practices of gender segregation have not prevented women from participating in the labor force, but they have prevented women and their families from admitting that they engage in such work. This has resulted in social justice initiatives using the volunteer label to employ women, but to do so under conditions that would normally be considered exploitative. Volunteers are not paid a salary but instead offered a meager stipend, if anything at all, to defray transportation and food costs. The ILO defines volunteers as "those of working age who [...] perform any unpaid, non-compulsory activity to produce goods or provide services for others."[7] Such a loose definition illuminates how gendered understandings of work and productive labor are constantly reproduced. Similarly, in Chapter 6, on youth migrants in Zimbabwe, Stanford Mahati tracks how gender norms render some forms of work invisible while others are made hyper-visible. These chapters elucidate how gender, religion, class, and ethnicity shape the categories of work and non-work.

The Global Polio Eradication Initiative is a non-state actor that uses the fungibility of the term "volunteer" to mobilize women to conduct the labor needed to achieve their goals. Closser's fieldwork in Pakistan also reveals the ways in which the philanthropic mission driving these programs naturalizes the predominantly female volunteer force's work as community participation and non-work. Social justice programs, with their reliance on underemployed local people to provide labor, help to entrench existing gender norms and persistent inequalities. Volunteerism becomes an alibi through which the global health industry takes "advantage of conditions of high unemployment to extract the labor of the vulnerable poor, particularly women, at low cost," Closser contends (see p. 58 in this volume). Indeed, the use of volunteers becomes a selling point to secure more donors.

In Chapter 7, Catherine Bowman tracks how a cultural program became a primary site from which labor categories for young people were transformed. Established during the height of the Cold War, the Summer Work Travel program allowed young people, designated as cultural ambassadors, to visit the United States during the summer and engage in work opportunities. In the following six decades, as the program's focus shifted to recruits from Western Europe and those from other parts of the world, the cultural ambassadors became part of the contingent labor force. Under the Summer Work Travel program, young people entered the US labor market with a J-1 visa. In order to ensure that these visitors' work conditions did not violate larger US labor laws pertaining to wages or work hours, the program had to categorize the young people as camp counselors and trainees. These new categories of "non-work" were created to "protect" US young people. The shifting definitions of this program are symptomatic of the difficulties of classifying work, non-work, employment, and unemployment. Chapter 7 also highlights how the laborscape in the United States is shaped by larger geopolitical factors, such as the collapse of the Soviet Union and economic transformations in Asia. It also maps the rhetorical shifts that allowed the program to become a primary focus for politicians' ire, a key node for discourse regarding the "theft of jobs" by foreigners that prevailed in the US 2016 presidential elections.

Shifts in the Argentine economy have resulted in a similar blurring of work/ non-work categories, making an even fuzzier demarcation between underemploy-ment and unemployment. Mariano Perelman (Chapter 5) raises the possibility that those who eke out a living picking through trash or selling goods piecemeal on the street might represent a different category of worker than those who are formally employed as sanitation workers or vendors at brick-and-mortar stores. Conducting fieldwork among street vendors and waste pickers (*cartoneros* and *cirujas*)—occupations that are stigmatized and often considered not to be work— Perelman highlights how, in a context of labor precarity, "the traditional limits between being employed and unemployed acquire new edges" (p. 76 in this volume). The meanings attached to these occupations shift depending on histor-ical and political contexts. By focusing on the voices of the "unemployed," Perelman is able to underscore the processes of social disaffiliation and down-ward social mobility.

While male young people predominated the ranks of the *cartoneros* and *cirujas*, women appeared almost always as pregnant subjects who continued to labor through their pregnancy. This underscores the gendered differences in the experience and phenomenon of workforce participation. Perelman concludes that, unlike their parents' generations—for whom work functioned as a crucial path in the transition from youth to adulthood, and in many cases also as the moral vector of social behavior—young people in Argentina currently lack "access to life." With persistent unemployment, jobs are no longer the traditional pathway to adulthood among young Argentinians, thus altering local and even national traditions and practices. Notwithstanding the tumultuous shift this entails for individual experiences of life, these young workers operating in the informal labor sector remain unaccounted for in economic and political data, reminiscent of Giorgio Agamben's (1998) evocative theorizing of bare life.

Carlo Inverardi-Ferri (Chapter 10) limns the contours of a new labor regime in post-Maoist Beijing. This ethnographic account of the gray economy details how rural migrants in Beijing cope with marketization processes and the concomitant transformations in social relations. Characterized as "fall-back work," Inverardi-Ferri's account reveals how waste recycling has become a way for young people to adapt their livelihood strategies to challenge key features of China's variegated capitalisms. The turn to non-waged work and informal circuits, Inverardi-Ferri contends, allows young people to regain control over labor time in the spheres of production and social reproduction; however, their passage to adulthood has become complicated as their lives are entangled in global and local dynamics. Central terms in this chapter—gray economy, informal work, and liminality—all hint at the lack of definitional clarity that we believe is central to the youth unemployment laborscape.

Together, these chapters address the ways in which the discourse about employment acts as a subterfuge against recognizing workers as such. Young people suffer the consequences of that flexible definition. These chapters investigate a central paradox concerning the dynamics of youth unemployment: some young people work far too much for little or no pay, while others are chronically jobless despite extensive training and exhaustive searches for suitable employment.

Migration

The connection between unemployment and migration is well established. Neoclassical economists see migration as an important factor in the international production system because it serves to adjust and regulate the supply and demand for labor. While this has changed to some extent as the industrial economy transformed itself to become service-oriented, it remains the case in the EU, where intra-EU labor migration is promoted as a way to adjust and ensure labor allocations within this economic region (O'Reilly et al. 2015). In the current situation of high unemployment, however, when the number of unemployed far exceeds the available jobs, labor migration can hardly fit the neoclassical economic models. Nevertheless, millions of young job seekers migrate regionally and internationally in order to escape lives of poverty, violence, and conflict. Current migration patterns vary by the education levels of the migrants, their social and cultural capital, and the level of economic development in both their country of origin and destination.

Young workers migrate from rural to urban areas, from one urban area to another, inter-regionally, and internationally. Motivated by severe push factors, unemployed and underemployed young people are willing to take great risks to seek employment, and in fact some of the figures suggest that "roughly 27 million leave their countries of birth to seek employment abroad as international migrants."[8] The geographical patterns of this current labor migration exhibit a move of southern Europeans to the United Kingdom, France, and Germany; of Irish people (Chapter 8) to China, Canada, and Australia; and of Africans, especially from sub-Saharan Africa (SSA) and North Africa, north to Europe. Even though the numbers of Africans who hope to get to Europe are still high, their numbers have declined as the former colonizers have enacted stringent immigration policies for subjects of their former colonies. Those who cannot reach Europe but strive for an international destination now favor China, the Gulf states, and the Americas. But the majority of African labor migrants, as Chapters 6 and 9 show, are intraregional.

The four chapters that pertain to the migration of unemployed young people (6, 8, 9, and 10) explicitly provide examples from three different continents, and three geographic settings and scales. Chapter 8 discusses Ireland, a country that has experienced several waves of emigration prior to the current one; Chapters 6 and 9 examine migration patterns of African unemployed young people; and Chapter 10 analyzes the phenomenon within marketized China. They explore the migration of unemployed and underemployed young people within the following scales:

- *inter*national, across political boundaries (Chapters 6, 8, and 9);
- *intra*regional, within an openly defined economic zone (as in the case of Chapter 9) or among countries that have signed bilateral trade agreements (as seen in Chapter 6); and
- *intra*state, from rural to urban areas (Chapter 10).

Beyond looking at the migration patterns of unemployed young people in these contexts, these chapters raise important issues about who migrates, the chosen destinations, the risks young migrants face, the impact of the influx of migrants on the sending society, and—at least in the context of Chapters 6 and 8—the discourses and representations of both youth unemployment and emigration at home (Chapter 8) and among those whose task is to help migrating unemployed young people (Chapter 6).

Who migrates?

All migrations, but particularly ones driven by poverty, produce anxieties. Unemployed and underemployed young people who are willing to travel dis-tances to improve their lives migrate to places where they seldom know or understand the laborscape. They take big risks, and this is particularly so for African migrants (Chapter 9). Sometimes migrants contract with smugglers who take their money and disappear or put them on overly filled boats, which sometimes capsize at sea. They risk being bought and sold in slave markets in Libya (Graham-Harrison 2017) or caught by border patrols and returned to places whose miserable situation they tried to escape in the first place. These migrants are desperate, like many in their communities of origin, but these young people are the ambitious, brave, and resourceful ones who believe they can endure the associated hardships.

Given the difficult journey that unemployed young people take, it is clear why most of the migrants are male, even though women's unemployment is often as high, if not higher, than that of men. Women are less mobile and are more likely to stay close to home, especially in traditional societies and when they have children. And if they do choose to migrate they often experience hardships that men do not, including sexual abuse and other forms of predation, especially when migration is less regulated, as in developing countries.

How do migrants choose their destinations?

Unemployed migrants include both educated and uneducated young people, and the levels of their education and skills often determine both the distance traveled and the choice of destination. In general, uneducated and low-skilled young people tend to migrate shorter distances. In the case of Africa (Chapters 6 and 9) they migrate *intra*regionally and in the case of China (Chapter 10) *intra*state. In these cases, they migrate from rural to urban areas and from smaller urban areas to larger ones. The educated unemployed young people tend to migrate internationally, as the case of Ireland (Chapter 8) shows. Regardless of their place of origin (i.e., Africa or Europe), the more educated young people often have the means, resources, social capital, and relative security to look for jobs that fit their educational levels; they may carefully evaluate their destination options, which is a luxury the unskilled poor simply do not have.

But if we look at the migration flows within the EU, especially from east to west in the period prior to the 2008 economic meltdown, we will notice that many young migrants, despite their high educational attainments, were so keen to leave their countries of origin that they found employment opportunities in low-skilled jobs with little to no possibility of growth (Kureková 2011 cited in O'Reilly et al. 2015: 6). Since the 2008–2009 economic crisis, however, migration of unemployed or underemployed young people from Central and Eastern Europe has slowed down, but return migration of newly unemployed young people who had previously found jobs in other parts of the EU has been on the rise (Zaiceva and Zimmermann 2014). Return migration has not been researched sufficiently to discern specific patterns.

In Africa, as Chapter 9 shows, education determines the distance and the type of migration that job seekers will take. Those who are educated usually stay put and wait until they can find a job that fits their skills. Their move is carefully calculated, and when it occurs it is usually with some knowledge of the laborscape at their destination. But this is hardly the case for the uneducated, low-skilled young people in Africa, those who constitute the majority of migrants in Africa. When they migrate, as Chapters 6 and 9 show, they migrate within the region. They follow historical migration patterns and take advantage of the economic agreements of West African states that were signed over three decades ago and have eased the movement of people and capital within the region.

Most migrations of skilled job seekers are within urban areas. Unemployed or underemployed unskilled young people, especially in the developing world, migrate mostly from rural to urban areas. Urban areas are attractive because of the high concentration of employment opportunities, and for entrepreneurial migrants the possibilities in the informal economy are infinite. In Chapter 10, Inverardi-Ferri shows that young, poor, rural migrants engage in a variety of livelihood strategies, mostly in the informal sector, because it affords them control over their labor time while providing them with a multitude of employment choices. His discussion of waste recyclers illuminates the resilience and resourcefulness of young waste pickers as they see trading in recyclables as an opportunity to gain more freedom and at the same time improve their material means. He further shows that, after a long period of rural to urban migration and involvement in the same "industry," some recyclers return from their visits home accompanied by younger siblings and family friends, creating a chain migration that further drains the poor rural areas of young people.

What are the risks and impacts?

For many young people, migration proves to be difficult regardless of their education and destination. The effects are felt on multiple levels and can have a serious impact on social integrity, family, community, and even the future of their countries of origin and destination. At the household level, the traditional or close-knit family is likely to disintegrate, especially if more than one child leaves for the city. This means that the already poor parents must fend for themselves unless they receive remittances, which for unemployed young people are hard to come by. At the level of the community in rural areas, the loss of entrepreneurial and resourceful young people can result in a generation gap, leaving the community imbalanced demographically. This includes the children who will now be born elsewhere, leaving the community at a deficit. Another side of the impact of emigration, however, especially if labor migrants are in a position to send remittances, is investment in projects that would not have happened through regular economic development but are now possible (Ajaero and Onokala 2013). At the national level, as Eleanor O'Leary and Diane Negra show in Chapter 8, the emigration of educated young Irish in search of jobs, in the aftermath of the 2008–2009 economic crisis, has been a cause for concern, although while it actually occurred little attention was given to it by politicians. Only when they realized the brain drain following the loss of the second children born to *émigrés*, and understood the magnitude of the loss of tax revenues, did Irish officials call for the young *émigrés* to return home.

Clearly, the impact of migration on places of origin, from where the ambitious unemployed or underemployed young people emigrate, requires careful consideration on the part of governments. At the very least, they need to enact policies that keep people in school, provide training, and ease the creation of new job opportunities that could absorb the sea of unemployed.

Policy

The laborscape of youth unemployment is the product of the contradictory dynamics of global political economy, but it is also the result of varied policy solutions that exist to challenge and shape it. Youth unemployment is a persistent and stubborn problem with no easy fixes. As Andrews and Prasch note in Chapter 2, market societies prefer "immediate policy problems and priorities" more than they do those concerning "the more distant future" (p. 25 in this volume). Today, volatile markets and unfavorable political circumstances have rendered long-term solutions even more elusive. In a world of extreme inequality, expensive higher education, and a rising retirement age, finding good work for young people is even more challenging. For these reasons, our book (especially in Chapters 3, 5, 7, 8, 10, 11, 12, and 13) explores important policy proposals that approach the problem.

The panoply of policy solutions that attend to the youth unemployment problem falls into two general typologies. Some regulate labor market conditions and imply a state intervention. Others regulate unemployment, or attempt to improve youth access to jobs. In other words, while the youth unemployment crisis is a particular problem requiring tailored solutions, the general approach to it is typical of other welfare-state provisioning.

There has long been an array of policy fixes dedicated to extending job training programs to fill what employers claim is a growing "skills gap," i.e., the situation in which the job market's needs are out of sync with the skills base of many high-school and college graduates.[9] These policies often note the persistence of a strange paradox: massive youth unemployment that far outnumbers the available jobs. One survey, for example, the Talent Shortage Survey conducted by Manpower Group in 2012, identified that 81% of employers in Japan, 71% in Brazil, 50% in Australia, and 48% in India reported difficulty when trying to match hires with their available jobs due to a skills mismatch. Policies directed against a skills mismatch are the most common way to support youth employment in sub-Saharan Africa (SSA), for example (Brooks et al. 2014). Other programs there are directed at wage earners, even though the majority of young Africans do not find entry-level work in the formal economy or in entrepreneurship, where low-interest loans are often unavailable for start-ups (Brooks et al. 2014). Wage subsidies that support employers looking to hire young people are not common in SSA except in South Africa, where a staggering 70% of black young people are unemployed (Betcherman and Khan 2015).

In Chapter 11, Crispen Chinguno dissects the arguments for and against the South African government's proposal to grant wage subsidies to employers who hire young workers. Because in South Africa unemployment among blacks is far greater than among whites, the subsidy policy put forth by the treasury could replace white workers with black ones and older workers with younger ones, worsening already-strained race relations. Advocates have argued the policy would intervene in an increasingly unregulated market to help a targeted group of young workers. The main trade union federation, by contrast, has proposed that the government should levy a tax against employers to gain a source of income to be paid directly to workers rather than drawing it from the state. Chinguno navigates this dilemma to discern lessons that will be useful in policy circles across the globe. One reason why the policy is so controversial is that the stakes are so high: South Africa has the third highest level of youth unemployment in the world. The country just above it, Spain, is located in the place we typically associate with the most generous policies to combat this crisis.

Cinalli and Giugni (2013) describe the emergence of what they call "youth unemployment regimes" in Europe: a panoply of policy instruments to address the variegated crisis as it manifests in different parts of the continent. Broadly, these differ to the extent they emerge from what Hall and Soskice (2001) call a "coordinated" or "liberal" market economy, signifying differences in the character of welfare state provision and capitalist development.

The Nordic countries, for example, face an aging population and a growing youth unemployment rate, and have sought to address these related problems by increasing the size of the workforce participation rate through their "Jobs for Everyone" initiatives, a series of different program types tailored to young workers. Under that umbrella, Iceland, which suffered the biggest shock of any Nordic country after 2008, has developed a series of guidance and motivational programs, and has even provided unemployment insurance[10] to students while they are studying in order to tighten the labor market and raise wages for those in jobs (Djernaes 2015). In Finland, Norway, and Sweden all unemployed young people are offered entrance into two-year job-readiness programs tailored to their specific interests and needs. Norway has a relatively different context owing to large oil reserves that provided a means to weather the financial crisis better than most. Still, relatively high youth unemployment rates persist there; some argue that overly generous welfare benefits, which enable young people to defer their entry to the labor market while still living comfortably, are the cause of those high rates (Djernaes 2015).

Based on this very assumption, Australia set its benefits packages for unemployed young people very low in order to make even low-wage work preferable to welfare access,[11] and set high punishments for failing to abide by the strict regulations of its NewStart program. There is some evidence, however, that this punitive principle, which has been combined with a series of undignified and useless jobs (bead sorting and hand-picking clover from public grass), has only served to punish more than employ (Bennet 2015). For example, Australian male youth unemployment, at 13%, is now higher than it was immediately following the 2008 crash.

But generous benefits do not keep youth unemployment artificially high in Germany, where significant levels of social welfare provision have been combined with a system of world-renowned apprenticeships. The German vocational training system facilitates transitions from school to work, which has kept its NEET population low even throughout the crisis. This system enjoins practical firm-based apprenticeships with school-based training, and creates strong links between schools and market demand. These programs combine an approach that starts students as early apprentices with businesses. Germany has promoted what it calls a "youth guarantee," and has pledged over US$10 billion to make sure every young European—including all eight million of the currently unemployed youth sector—has a job. Theoretically, that money should flow out to the places hardest hit by the crisis, but Spain is evidence that it is not happening yet.

In Spain, despite the fact that the right to work is enshrined by constitutional decree, over half of the country's young people cannot find a job—in 2016 approximately half of the country's young people were out of work. Ciro Milione (Chapter 12) describes the varied legal and policy attempts and failures to combat this problem in the hard-hit region of Andalusia. Two such programs, the "Programa Emple@Joven" and the "Iniciativa @mprende+" for example, allegedly guarantee a job offer to recent graduates. But the programs are too new to generate the data necessary to understand their impacts. The youth unemployment rate remains stubbornly high in places where these programs have been operating for a number of years, which leads Milione to voice some degree of pessimism.

As O'Leary and Negra demonstrate in Chapter 8, Ireland is another EU outlier in terms of youth unemployment: the country seems to have quietly but quickly addressed the crisis in the wake of the 2008 recession by "disappearing its young people" in a massive exodus unmatched throughout Europe. While this can hardly be regarded as official policy, O'Leary and Negra demonstrate that the trend was inspired by domestic policy intended to bail out the banking sector in the wake of the crisis, which was seen as disastrous by the nation's young people, and can be considered a push factor driving out-migration.

Outside of Europe the laborscape changes again. Japan, for example, has an enviably low unemployment rate, but Gottfried makes clear in Chapter 3 that the precarious nature of jobs complicates the problem nonetheless. Japan's "Revitalization Strategy" set a goal of reducing the number of unemployed for six months or more by 20% in order to reach an employment rate of 78%.[12] One way it has sought to do this is by encouraging entrepreneurship[13] among the young, by helping them start crowd-funding programs and by providing low-interest loans to young business developers. Japan's "Industrial Competitiveness Action Plan" also calls for strengthening ties between industry and education by fostering contact between young students and workplace leaders as early as elementary school.[14] Finally, active labor market policies include the "Vocational

Training with Practical Work" initiative, which subsidizes businesses that train young workers for a period of six months to two years.[15]

Elsewhere in Asia, as Inverardi-Ferri explains in Chapter 10, waste recyclers in China amplify some of the main challenges facing policymakers there. The large informal sector of young male migrants now represents just over half of the working young people in major cities. The main policy approach to this problem is to increase pupil-to-business connections within migrant schools, yet this has been frustrated by massive social discrimination against migrants. Other reliable data on the political approach to the problem are scarce, but the ILO (2014) reports that entrepreneurs have access to low-interest government loans and tax rebates when starting ventures. Moreover, while child labor still persists in rural areas, progress has been made to eradicate it in major cities, yet waste recycling is one area where it still flourishes.

Waste recyclers and other informal street workers populate Perelman's narrative in Chapter 5 on Argentina, where neoliberal reforms have helped to transform the political economy and obscure the very nature of what it means to work. He theorizes the "edges" of this category as it relates to the liminal state in which many informal workers find themselves. Nearby Brazil may offer a useful counterexample. Lula da Silva's presidency—from 2003 to 2010—deployed an aggressive combination of social welfare schemes aimed at reducing inequality, raising wages for working-class families, and expanding public financing for education and healthcare. These provisions were conceived as broad social welfare policies, but they had important impacts on curtailing the youth unemployment problem, even as it existed for a large population of the informal sector. The formal sector has since expanded, providing more inroads for young workers into jobs with labor rights and regularized wages.

Brazil's experience suggests that expanding labor rights is one strategy to curtail youth unemployment, a counterintuitive finding given the assumptions of most policymakers who fear that unions will drive up wages, drive out businesses, and make it more expensive to hire new people. But research shows that "the key to reform" in low-wage work "lies in enabling young, first time workers to stand up for their rights" (Santos and Randolph 2014). The masses of unemployed young people worldwide appear to embody many of the characteristics—insecurity, flexibility, high mobility, and poverty—that apply to what Standing (2014) dubs the "precariat." This "new dangerous class" is, according to the theory, likely to embrace nontraditional forms of labor organization in order to assert itself, and eschew the bureaucratic top-down character of trade unionism; but there is limited evidence that this has happened. Nonetheless, the model of union and community partnerships, which have been common for decades in the United States and Latin America, might be spreading. The 2011 uprising in Egypt, a movement led by young people, helped to spawn independent trade unions in its wake to promote jobs and worker rights. Today, Egyptian young people lack job opportunities in the wake of the turmoil and, where jobs do exist, the precarious nature of employment makes those jobs wildly insecure and low paying (Kindt 2014).

The laborscape in the United States, where male youth unemployment is more than double the overall rate, is the result of conflicts stemming from disparate pieces of the varied dynamics described above. Since the crisis of 2008–2009, the most robust policy proposal to tackle unemployment is work sharing. As it sounds, the policy reduces the average number of hours that each employee works to avoid layoffs or to increase hires. First used during the Great Depression to "spread the work," the idea has had occasional success in the United States, although Germany has used it especially well. Vroman and Brusentsev (2009) have argued that a particular kind of short-term compensation program, which is mainly used during especially long and troubling times of youth unemployment, could "preserve the diversity of a company's workforce" by retaining employees who are usually the first to be let go, including young people. While the program is not especially directed at young people, given that they lack seniority (one rationale for early firings), this program has the potential to disproportionately benefit young workers (Maich et al. forthcoming). Moreover, thanks to an Obama-era revival, it already exists. About half of US states have an active work-sharing program that allows firms to apply for a subsidy to retain workers at fewer hours without a reduction in pay, thereby allowing them to keep other workers employed.

While changing the duration and tempo of the working week is one strategy proposed to positively effect youth employment, other experts have suggested changing the parameters that define the lifetime of a worker. Would lowering the retirement age make room for younger workers at the other end of the life-course work span? While lower retirement ages are correlated with lower youth unemployment rates in Europe, there does not seem to be a causal mechanism to ensure that the two dynamics are necessarily interrelated, even in the global south (Munnell and Wu 2013).

The presence of migrant workers in the United States is consistently used as an explanation for stubborn unemployment rates. With talk of an expanded border wall at the forefront of these debates today, intense focus has been put on the policies that force young workers from outside the country to seek employability in the United States. In Chapter 7, Bowman focuses on one such longstanding policy that usually goes unnoticed but is crucial to understanding this dilemma: the J-1 Summer Work Travel program. Although the program has been a contentious flashpoint for varied arguments about migrant workers, the cost of labor, and even the very nature of what it means to be a worker, there are not enough data available to evaluate the program. She therefore makes an explicit demand for access to the resources that will help to craft the most effective policies.

Many experts routinely encourage US policymakers to explore the kinds of internship and apprenticeship programs that are common in Germany. Links between schools and employers are critical determinants of youth employability. Some in the United States have made a farcical interpretation of these data, and pursued policies to transform schools into little more than workplace training centers. The sensible policy goal, however, is to connect one sphere to another, not to assimilate them. Connecting students with jobs also means disconnecting students from prisons.

Many commentators have found the schools-to-jobs pipeline too "leaky," with recent graduates taking long detours in unemployment. By contrast, in Chapter 13, Marcellus Andrews sees the school-to-prison pipeline as all too direct, and finds controversial reasons for why this might be the case. Rather than focusing explicitly on anti-black racism, his "stone-cold materialism" as an economist calls attention to two dynamics. First, he notes the unintended consequences of bad choices made by young workers as they navigate a complex world in which they are challenged to find housing, for instance, or tempted to experiment with drugs or resort to crime. Second, and most importantly, he criticizes the penchant for punishment by political elites over other prescriptive solutions. Ultimately, his chapter implies a call for resistance led by those most impacted by carceral society.

The economic crisis drove youth unemployment rates to shocking levels, and the long recovery over the past decade has meant a respite from the worst conditions. But the persistence of the problem is a testament to the need for more creative and robust policy strategies, not political partisanship.

Concluding remarks

The collected chapters here establish a new understanding of the depths of the youth unemployment crisis. While not an exhaustive list, the book illuminates an inchoate laborscape in the core and peripheral regions of the world economy, and establishes links between them. Youth unemployment appears to be both a symptom and cause of the varied challenges facing young people today. And while there are no easy solutions, we are heartened by the fact that experiments to ameliorate this crisis are varied and ongoing. We hope this volume offers both scholars and lay audiences a window into this problem so that we may better understand it and someday solve it.

Notes

1 While originally produced in the United States, versions of the show were aired in 17 other countries. It proved to be a hit formula everywhere but Germany and Finland, where participants were asked to leave the show rather than being told they were fired.

2 The ages included in the category "youth" vary widely across data sources. When we deploy diverse data sources we indicate their definition of "youth" rather than trying to produce a singular age group. This particular figure was taken from the most currently available data at the ILO: www.ilo.org/global/topics/youth-employment/lang–en/index.htm (retrieved April 29, 2018).

3 www.un.org/youthenvoy/2015/10/envoy-youth-global-youth-economic-opportunities-summit (retrieved January 26, 2018).

4 www.ilo.org/wesodata/?chart=Z2VuZGVyPVsiTWFsZSIsIkZlbWFsZSJdJnVuaX Q9IlJhdGUiJnNlY3Rvcj1bIkluZHVzdHJ5IiwiU2VydmljZXMiLCJBZ3JpY3VsdH VyZSJdJnllYXJGcm9tPTE5OTEmaW5jb21lPVtdJmluZGljYXRvcj1bIn

VuZW1wbG95bWVudCJdJnN0YXR1cz1bIlRvdGFsI0mcmVnaW9uPVsiV29ybG
QiXSZjb3VudHJ5PVsiSW5kaWEiXSZ5ZWFyVG89MjAxOSZ2aWV3 3R
m9ybWF0PSJDaGFydCImYWdlPVsiQWdlMTVfMjQiXSZsYW5ndWFnZT0iZW4i
(retrieved April 29, 2018).

5 www.hrw.org/world-report/2017/country-chapters/dangerous-rise-of-populism (retrieved June 25, 2018).

6 www.ilo.org/ilostat-files/Documents/description_UR_EN.pdf (retrieved April 29, 2018).

7 www.ilo.org/global/statistics-and-databases/statistics-overview-and-topics/ WCMS_470308/lang–en/index.htm (retrieved April 29, 2018).

8 www.ilo.org/wcmsp5/groups/public/—ed_emp/documents/genericdocument/ wcms_209613.pdf (retrieved January 21, 2018).

9 Skills development programs are among the most common policy solutions across the world, even despite growing evidence that a skills gap does not exist to the extent we typically think it does (Weaver and Osterman 2016); however, so much policy is predicated on the assumption that it does exist and is therefore important to describe.

10 www.elmmagazine.eu/articles/a-nordic-perspective-on-youth-unemployment (retrieved April 29, 2018).

11 www.acoss.org.au/acoss-employment-proposals-for-2015-federal-budget (retrieved April 29, 2018).

12 www.ilo.org/wcmsp5/groups/public/—asia/—ro-bangkok/—sro-bangkok/documents/ publication/wcms_534277.pdf (retrieved April 29, 2018).

13 www.ilo.org/dyn/youthpol/en/f?p=30,850:1201:0::NO::P1201_RESPONSE_SE T_ID:226 (retrieved April 29, 2018).

14 www.kantei.go.jp/jp/singi/keizaisaisei/pdf/housin_gaiyou_140124en.pdf (retrieved April 29, 2018).

15 www.ilo.org/dyn/youthpol/en/f?p=30,850:1201:5,849,776,079,686,679::NO:1201: P1201_RESPONSE_SET_ID:300 (retrieved April 29, 2018).

References

Agamben, Giorgio. 1998. *Homo Sacer: Sovereign Power and Bare Life*. Stanford, CA: Stanford University Press.

Ajaero, Chukwuedozie, and Patience C. Onokala. 2013. "The Effects of Rural-Urban Migration on Rural Communities of Southeastern Nigeria." *International Journal of Population Research*. Retrieved from www.hindawi.com/journals/ijpr/2013/610193 (accessed February 25, 2018).

Alexander, Michelle. 2012. *The New Jim Crow: Mass Incarceration in the Age of Colorblindness*. New York: The New Press.

Appadurai, Arjun. 1996. *Disjuncture and Difference in the Global Cultural Economy*. Minneapolis, MN: University of Minnesota Press.

ASDA'A Burson-Marsteller. 2017. "Arab Youth Survey: The Middle East, A Region Divided." Retrieved from www.arabyouthsurvey.com/about-survey.html (accessed February 15, 2018).

Baah-Boateng, William. 2016. "The Youth Unemployment Challenge in Africa: What are the Drivers?" *The Economic and Labour Relations Review*, 27(4): 413–431.

Beck, Ulrich. 1992. *Risk Society: Towards a New Modernity*. Thousand Oaks, CA: Sage. First published 1989.

Bennet, Owen. 2015. "Our Punitive Welfare Policies Crush the Unemployed and Keep Them Out of Work." New Matilda.com. Retrieved from https://newmatilda.com/2015/11/

20/our-punitive-welfare-policies-crush-the-unemployed-and-keep-them-out-of-work (accessed April 14, 2018).

Betcherman, Gordon, and Themrise Khan. 2015. *Youth Employment in Sub-Saharan Africa: Taking Stock of the Evidence and Knowledge Gaps*. Ottawa, ON: International Development Research Centre.

Brooks, Karen, Deon P. Filmer, M. Louise Fox et al. 2014. "Youth Employment in Sub-Saharan Africa." Washington, DC: World Bank. Retrieved from http://documents.world bank.org/curated/en/784471468000272609/Overview (accessed April 15, 2018).

Card, David. 2011. "Origins of the Unemployment Rate: The Lasting Legacy of Measurement without Theory." Retrieved from http://davidcard.berkeley.edu/papers/origins-of-unemployment.pdf (accessed April 24, 2018).

Cinalli, Manlio, and Marco Giugni. 2013. "New Challenges for the Welfare State: The Emergence of Youth Unemployment Regimes in Europe?" *International Journal of Social Welfare*, 22: 290–299.

Djernaes, Lars. 2015. "A Nordic Perspective on Youth Unemployment." *Elm Magazine*. Retrieved from www.elmmagazine.eu/articles/a-nordic-perspective-on-youth-unemploy ment (accessed April 14, 2018).

Foucault, Michel. 1970. *The Order of Things*. New York: Vintage. First published 1966.

Graham-Harrison, Emma. 2017. "Migrants from West Africa Being 'Sold in Libyan Slave Markets.'" *Guardian*, April 10.

Hall, Peter, and David Soskice. 2001. *Varieties of Capitalism: The Institutional Foundations of Comparative Advantage*. Oxford: Oxford University Press.

Hochschild, Arlie. 1989. *The Second Shift*. New York: Penguin.

Idris, Iffat. 2016. *Youth Unemployment and Violence: Rapid Literature Review*. Birmingham: GSDRC.

International Labour Office (ILO). 2014. *Youth Employment Policy Brief: China*. Bangkok, Thailand: Asia and Pacific Youth Employment Programme.

Kessler-Harris, Alice. 1982. *Out to Work: A History of Wage-Earning Women in the US*. New York: Oxford University Press.

Kindt, Kristian Takvam. 2014. "The New Egyptian Labor Movement. Giving Youth a Voice in the Workplace." JustJobs Network. Retrieved from http://justjobsnetwork.org/wp-content/pubs/reports/FAFO.pdf (accessed April 29, 2018).

Kureková, Lucia. 2011. *The Effects of Structural Factors in Origin Countries on Migration: The Case of Central and Eastern Europe*. International Migration Institute, Working Paper 45. Oxford: Oxford University Press.

Maich, Katherine, Jamie K. McCallum, and Ari Grant-Sasson. Forthcoming. "Chapter 10: Time's Up! Shorter Hours, Public Policy, and Time Flexibility as an Antidote to Youth Unemployment." In *Youth, Jobs and the Future: Problems and Prospects*. L. Chancer and M.S. Jankowski, eds. Oxford: Oxford University Press.

Manpower Group. 2012. "How Policymakers Can Boost Youth Employment." Retrieved from http://files.shareholder.com/downloads/MAN/2113234660x0x600960/1f9d24d2-d737-40c7-805f-a38b183e7766/How%20Policymakers%20Can%20Boost%20Youth%20Employment%20FINAL%2009-18-12.pdf (accessed April 15, 2018).

Marx, Karl. 1976. *Capital: A Critique of Political Economy*, Vol. 1. London: Penguin Harmondsworth. First published 1867.

Munnell, Alice H., and April Yanyuan Wu. 2013. "Do Older Workers Squeeze Out Younger Workers?" Working Paper No. 13-011. Stanford Institute for Economic Policy Research (SIEPR). Retrieved from https://siepr.stanford.edu/research/publications/do-older-workers-squeeze-out-younger-workers (accessed October 15, 2017).

Nordås, Ragnhild, and Christian Davenport. 2013. "Fight the Youth: Youth Bulges and State Repression." *American Journal of Political Science*, 57(4): 1–15.

Nussbaum, Emily. 2017. "The TV That Created Donald Trump." *New Yorker.* Retrieved from www.newyorker.com/magazine/2017/07/31/the-tv-that-created-donald-trump (accessed April 24, 2018).

O'Reilly, Jacqueline, Werner Eichhorst, András Gábos et al. 2015. "Five Characteristics of Youth Unemployment in Europe: Flexibility, Education, Migration, Family Legacies, and EU Policy." Sage Journals. Retrieved from http://journals.sagepub.com/doi/full/10.1177/2158244015574962 (accessed January 10, 2018).

Oner, Ceyda. 2017. "Unemployment: The Curse of Joblessness." Retrieved from www.imf.org/external/pubs/ft/fandd/basics/unemploy.htm (accessed April 24, 2018).

Osakwe, Chukwuma. 2013. "Youth, Unemployment and National Security in Nigeria." *International Journal of Humanities and Social Science*, 3(21): 258–268.

Santos, Alfredo, and Gregory Randolph. 2014. "Brazil's Social Welfare Approach: Improving Job Outcomes for the Youth by Delaying Entrance into Workforce." In *Overcoming the Youth Employment Crisis: Strategies from Around the Globe*. Geneva: JustJobs Network.Retrieved from http://justjobsnetwork.org/brazils-social-welfare-approach-improving-job-outcomes-for-the-youth-by-delaying-entrance-into-workforce (accessed April 14, 2018).

Standing, Guy. 2014. *The Precariat. The New Dangerous Class*. London: Bloomsbury Academic.

Vroman, Wayne, and Vera Brusentsev. 2009. *Short-Time Compensation as a Policy to Stabilize Employment*. Newark: The Urban Institute, University of Delaware. Retrieved fromwww.urban.org/sites/default/files/publication/30751/411983-Short-Time-Compensation-as-a-Policy-to-Stabilize-Employment.pdf (accessed October 27, 2017).

Weaver, Andrew, and Paul Osterman. 2016. "Skill Demands and Mismatch in US Manufacturing." *ILR Review*, 70(2): 275–307.

Williams, Raymond. 1977. "Structures of Feeling." In *Marxism and Literature*. London: Oxford University Press, pp. 128–135.

Zaiceva, Anzelika, and Klaus F. Zimmermann. 2014. "Migration and the Demographic Shift." Discussion Paper No. 8743. Bonn, Germany: IZA and Bonn University.Retrieved from http://ftp.iza.org/dp8743.pdf (accessed February 22, 2018).

2 Youth unemployment and unnecessary intergenerational conflict

Marcellus Andrews and Robert Prasch

In most societies the first rule of economic life is very simple: no work, no income. Unemployment matters in market societies for the simple reason that most people earn their daily bread by using the only asset they possess: their capacity to work for someone else, usually for an organization owned by other people, too often a small coterie of very wealthy families. Mass unemployment, where a substantial portion of the workforce is jobless and will remain without work for quite some time, is an unfortunate but regular feature of capitalist economies. The beating heart of that system—banking and financial markets, including housing markets—suffers periodic breakdowns akin to heart attacks and strokes that paralyze the process of buying and selling, and which inexorably kill jobs and destroy lives.

The recent global breakdown of capitalism, evident in the crash that led to the so-called Great Recession (2007–2009), was the result of a toxic mix of lax government regulation, excessive financial speculation, runaway economic inequality, and forms of financial innovation that literally created fictive wealth. When the crisis hit, all that phony wealth on bank balance sheets could not be priced and therefore had no value, leaving banks without any way of knowing what they really owned or owed, and thereby impairing lending and spending across the globe. As any historically literate economist knows, not least the late Robert Prasch of Middlebury College, lightly regulated lending combined with stagnant labor incomes and an aggressive commitment by governments to bail out bankers is a recipe for system failure.

But there are many different types of unemployment affecting different categories of persons in very different ways. Cyclical unemployment occurs when an economy experiences a collapse in spending, as vividly demonstrated by the Great Recession. All workers, young and mature alike, suffer when cyclical unemployment increases, although the young suffer more for reasons explored below. Structural unemployment, on the other hand, is joblessness that occurs when a working population is too costly to employ or is unemployable because it lacks the necessary skills or because technological change renders formerly productive people obsolete. Trade and technology can turn a competent worker into a valueless person by replacing domestic labor with much cheaper foreign labor or by acquiring machines that necessitate fewer workers or workers with different sets of skills. Poorly schooled young people "graduating" from substandard schools or dropping out of school are far more likely to be structurally unemployed because they do not have skills the labor market values.

Young and in trouble

Youth unemployment—by which we mean joblessness experienced by young men and women ranging in age from the cusp of 20 to their mid-20s, who are either new to the labor force after graduating school or have only been available for work for a short time by virtue of their youth—is a distinct species of joblessness combining cyclical and structural features that present unique challenges to governments and communities. Unemployment among these young people can destroy their future by impeding their transition from childhood dependency to economic maturity. The Nobel laureate Amartya Sen reminds us that the purpose of economic life is to promote human freedom through the creation of realistic opportunities for men and women to develop their "capabilities": those capacities to act on their chosen life plans in pursuit of objectives that society sees as reasonable and valuable (Sen 2009). A community and economy that thwarts this transition from economic immaturity to economic autonomy is not only undermining the liberty of its youthful citizens but also tearing at the fabric of the social contract. A healthy democracy ensures that the young can grow into fully competent actors able to support themselves and their children while exercising their powers as citizens. High and enduring rates of youth unemployment, particularly structural unemployment due to technological change and persistent barriers to educational opportunity, are a sign that a society has reneged on its obligation to create a sustainable path from childhood dependency to economic maturity.

In this chapter we explore three of the more important aspects of the youth unemployment problem in the United States, the first being the economic logic of youth unemployment through the lens of basic economic theory. One particularly important source of trouble is that employers, as buyers of labor time, view young workers as less reliable and therefore less valuable than mature workers, with the consequence that job offers to the young pay less and are less reliable than those to older workers, even in boom times. The analysis of imperfect information in market economies combined with insights from basic game theory go a long way towards explaining why young workers have so much trouble in job markets.

Second, young workers are inadvertently harmed by economic policies that make it harder for them to get and keep jobs because governments ultimately cater to the economic needs of older labor (i.e., the voting majority), over the needs of younger labor. The modern welfare state can, under certain circumstances, offer older workers important economic protections that also deprive the young of work by making young workers too expensive. Governments must be careful to design their social benefit policies in ways that do not exacerbate this basic conflict between younger and older workers.

Third, youth unemployment is, to some extent, a feature of democratic market economies that display weak forms of social solidarity across generational and racial/ethnic lines. The needs of the young will always matter less than the needs of politics for the simple reason that mature and older voters outnumber the young, who in any case tend not to vote. Market societies and democratic polities, however, tend to favor immediate policy problems and priorities more than those of the more distant future, particularly in matters of schooling and redistributive strategies that promote the economic wellbeing of future populations, especially if these populations comprise racial and ethnic outcasts. The primary problem here is that the incentive structure of electoral politics in a society marked by low degrees of intergroup and intergenerational altruism will systematically bias fiscal policies towards mature and elderly middle-class populations over the young and poor—thereby leaving a portion of each rising generation without the skills and knowledge required to compete for jobs in the most dynamic sectors of the economy. These incentive problems create a sizeable economic underclass that shuttles between the low-wage labor market and prisons, a caste that "works" in the illegal sectors of the system and whose children are guaranteed to share their parents' life at the bottom of society.

A few "stylized facts"

The job market woes of young workers are nicely illustrated by looking at the statistical relationship between economic growth and unemployment known as Okun's law, named after the macroeconomist Arthur Okun, who first formulated the connection (Okun 1962). It should be noted that our purpose here is limited to characterizing the connections between unemployment and growth as revealed by macroeconomic data, not to pursue a detailed econometric study of how unemployment and economic growth are connected, despite the importance of that exercise.

Figure 2.1 shows the evolution of unemployment rates for US workers between the ages of 20 and 24 as well as for workers aged 25 and older between 1948 and 2014. Figure 2.1 confirms the well-known fact that young workers face much higher unemployment rates than older workers, and also experience greater fluctuations in employment prospects over the business cycle than their older counterparts.

Table 2.1 shows the average unemployment rates for younger and older workers since 1990, when the rate of growth of real GDP was above and below the historical average of 2.4% per annum between 1990 and 2016, as well as for years when GDP was falling and the economy was officially in recession. As one might expect, the table indicates that unemployment rates for young workers are not only higher than those for older workers under all circumstances, but variations in unemployment rates for young workers are much greater than for older workers. The trade-off between growth and changes in unemployment for older workers is much less pronounced than that for younger workers: economic expansion and contraction have a much smaller effect on older workers than

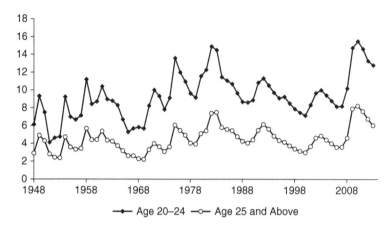

Figure 2.1 US unemployment rates by age (1948–2008) (Federal Reserve economic data, Saint Louis Federal Reserve Bank; adult rate: www.fred/stlouisfed.org/ UNRATE; youth rate: www.fred.stlouisfed.org/series/SLUEM1524ZSUSA)

Table 2.1 The relationship between real GDP growth and unemployment rates for younger and older workers (1990–2016)

GDP Growth Rate	Older Workers	Younger Workers
Above 2.4%	5.8%	12.0%
Below 2.4%	6.7%	13.6%
Below 0%	7.4%	14.6%
Maximum	4.0%	9.2%
Minimum	9.8%	18.4%

Source: Adult unemployment rate: www.fred.stlouisfed.org/series/UNRATE; youth unemployment rate: www./fred.stlouisfed.org/series/SLUEM1524ZSUSA; real GDP: www.fred.stloius.org/series/GDPC1 and authors' calculations.

younger workers. This fact, when combined with the information contained in Figure 2.1, not only indicates that younger workers experience higher unemployment rates than their older counterparts, but also that younger workers act as a "shock absorber" for older workers, bearing a greater burden of any economic downturn than the rest of the workforce.

These facts have a few unfortunate implications for the economic wellbeing of young workers, particularly under contemporary economic and political conditions. First, the US economy is a mature system that is unlikely to ever return to the halcyon days of high growth when the system could be counted on to expand at a 3% and 4% rate per annum, a rate associated with the "golden age" of American growth, from the end of World War II through the late 1970s. The post-World War

II boom that still lingers in the minds of politicians and the public was a special period of US dominance, one in which a combination of historic circumstances, technological changes, and a national commitment to full-employment economic policy combined to create a "high pressure economy" that kept GDP growth rates high and unemployment rates low across the age spectrum. US economic policy has long since turned away from promoting high levels of employment through active government management of aggregate demand, whether by sanctioning high levels of government spending financed by progressive taxation, a reasonable approach to the management of the public debt that allowed government to expand that debt in line with the growth of the economy, or a combination of both. The victory of US conservatism—a peculiar amalgam of deep animus against progressive redistribution in economic policy in favor of regressive redistribution, which itself promotes low taxes and limited regulation—combined with a blithe disregard for the once-hallowed conservative doctrine of "sound finance," all in the hope of inspiring growth by reducing the burden of taxation on the already rich, means that the average growth rate in the United States is likely to stay low for the foreseeable future. This shift from "demand side" economic policies to "supply side" policies seeks to boost the profitability of investment in new capital to spur economic expansion; since it began in the 1980s it has met with little success in matters of employment for all, and especially young workers. There is little if any evidence that reducing taxes on the very wealthy creates jobs for anyone (Slemrod and Bakija 2008). Of course, the failure of the supply-side approach to economic policy is well documented. But one consequence of the now decades-long program of regressive redistribution, in the hope of inspiring capital to grow, is that a primary source of jobs for young people—a high level of demand for labor in a "high pressure economy"—is foreclosed.

One aspect of recent US economic experience, however, seems to favor young workers over older workers: the growth of wages over the past 25 years. Table 2.2 shows the rate of growth in the average inflation-adjusted earnings of workers by age and highest level of education attained between 1991 and 2016, the latest period for which data are available. Young workers during this period have experienced faster earnings growth than their older counterparts across the educational spectrum, with the exception of workers whose education ended after earning a high-school degree.

These data are enlightening because younger workers in 2016 earned more relative to their older counterparts than they did a quarter of a century ago (see also data from Table 2.3, which compare the inflation-adjusted average earnings of workers by age and education in 1991 and 2016). The fact that the young earn more relative to the old, despite the fact that younger workers experience higher rates of unemployment, suggests that the US economy in 2016 offered young workers more opportunities than in 1991. Indeed, Table 2.3 suggests that there was an element of earnings compression between the young and the old in the period from 1991 to 2016 (shown in the last two columns of the table).

The substantial decline in the ratio of wages between older and young college graduates is consistent with the observation that new and more effective

Table 2.2 The rate of growth of average earnings by age and education (1991–2016)

Education	Older Workers	Younger Workers
Less than 9th Grade	1.40%	1.72%
High School	0.68%	0.51%
Some College	0.29%	1.33%
Associate's Degree	0.22%	0.86%
Bachelor's Degree	0.84%	1.87%

Source: US Census Bureau, Historical Tables: People, Table P-28: Educational Attainment—Workers 25 Years Old and Over by Median Earnings, www.census.gov/data/tables/time-series/demo/income-poverty/historical-income-people.html.

Table 2.3 Average earnings by age and education, 1991 and 2016

Education	Young Workers		Older Workers		Ratio 1991	Ratio 2016
	1991	2016	1991	2016		
<9th Grade	$15,032	$23,145	$24,263	$34,478	1.61	1.48
High School	$19,802	$23,503	$41,149	$46,833	2.07	1.99
Some College	$13,601	$18,986	$49,035	$52,852	3.60	2.78
Associate's	$21,324	$26,681	$52,726	$55,795	2.47	2.09
Bachelor's	$26,339	$42,136	$78,210	$96,670	2.97	2.29

Source: US Census Bureau, Historical Tables: People, Table P-28: Educational Attainment—Workers 18 Years Old and Over by Mean Earnings, Age and Sex, 1991 to 2016, and authors' calculations.

technologies are embodied in the form of new workers; their relative scarcity (compared to incumbents embodying older forms of knowledge) results in wage premiums paid to younger workers in the most technologically dynamic parts of an economy. Although incumbents have reaped considerable benefits from having more experience than their younger counterparts—from their frequent participation in the design and development processes that generated new technologies, and from knowledge passed on in schools—the arrival of a new batch of well-schooled workers in the fastest growing parts of the system may be accompanied by rising earnings in this segment of the labor market.

Economic theory and youth unemployment

Young workers pose a distinct set of challenges for employers. First, they are usually less productive than their older counterparts, particularly in sectors characterized by mature technologies that are largely immune to substantial changes in educational credentials or knowledge requirements. Employers spend a great deal of time and effort in turning a young, inexperienced, low-productivity employee into an effective labor asset. Older workers in these sectors have more experience and therefore greater value to employers than their younger counterparts, at least in terms of direct labor costs. Lindbeck and Snower (1989) have shown that productivity differences between young and older workers ultimately depend on the willingness of mature incumbents to train their new colleagues, thereby granting incumbent workers significant monopoly power in dispensing knowledge and cooperating with new arrivals. When older workers are disinclined to integrate new workers into the work process, incumbents can put the young at a cost disadvantage, thereby protecting themselves from the potential competition of these new workers. Moreover, incumbent older workers currently holding jobs will give priority to their own interests over those of unemployed workers, including younger workers, with the consequence that wage bargaining between employers and incumbent workers can result in situations where the young and unemployed remain jobless because of deals made between older workers and their bosses.

The second challenge, directly relating to the productivity advantages of older workers, concerns a demand-side problem that is made worse by the inevitable risk associated with any hiring decision. Labor markets are, by their very nature, riddled with important information problems that have a profound effect on wages and employment for all workers, no matter their level of education, age, or work experience (Akerlof 1984; Bowles 2004: 167–198). Economists have developed an entire branch of economic theory— the economics of information— to assess the effect of knowledge and the lack thereof on economic wellbeing, including the way in which labor markets work (Stiglitz 2001: 479–481; Bowles 2004: 167–198).

Whenever a buyer and seller meet in the marketplace, each side makes a decision to engage in or refrain from a trade based on what he or she knows, whether about the commodity being exchanged or the reputation of the buyer or seller. In the labor market, an employer offers a job to a worker based on what can be known about that worker's skills, reliability, and demeanor, in just the same way that a worker will assess a job based on the employer's reputation for honesty and fairness, the state of working conditions within the enterprise, and other factors including the wages on offer. Most jobs are, in fact, investments wherein an employer chooses a worker who will then be trained at the employer's expense on the gamble that this capital outlay will be more than covered by the value of the product. The risk to employers is reduced when a worker's *résumé* indicates both a record of prior success and the appropriate educational credentials. Young workers are therefore at a disadvantage in established sectors of the economy because they lack experience, which makes them less productive and also a greater risk in terms of "paying back" the costs of job training.

The third risk associated with employing young workers is exacerbated by another fact of life: young workers tend to quit jobs at higher rates than their older counterparts, not least because the young are a bit more impulsive than older people whose forbearance is the product of realism born of experience. Employers may be more than willing to offer young workers jobs if the wage gap between mature and young workers is big enough to offset the risks of hiring the young. But the productivity advantage of older workers means that some very effective young workers will lose out if employers engage in this sort of statistical discrimination against the young in favor of the old and judge individual young workers on the basis of group averages. If a capable young worker is able to make his or her way up the promotion ladder over time, both by becoming older and by having success in the workplace, then he or she will in time *benefit* from discrimination against the young.

The simple economics of information suggests further barriers to the prospects of young workers. An unemployed young worker with a record of joblessness or a delay in entering the workforce is, from the perspective of an employer, a riskier hire than someone with a steadier work history; such a history thus blocks the road to employment for young workers unlucky enough to suffer early periods of joblessness. In turn, the damaging signal connected to being young and jobless leads to the phenomenon of "labor market scarring" for young workers: the well-known and harsh reality that early bouts of unemployment for young workers lead to lower lifetime employment prospects and earnings in the future. Over time, these early disadvantages lead to long-term troubles because a work history marred by a record of limited success proves to be a barrier to future success. Indeed, the empirical literature suggests that unemployed young workers in the United States can pay a lifetime wage penalty of up to 20% of the average wages of their cohort when their entry into employment is delayed, with the penalty rising with the length of their initial absence from work (Kahn 2010).

Elementary economics and unpleasant arithmetic

Our analysis suggests that young workers face two problems in modern labor markets. First, the bias of neoliberal economic policy towards tight budgets, low taxes, and limited government intervention puts a downward bias on aggregate demand, thereby limiting employment opportunities for the young. Second, the productivity and experience deficit of younger *vis-à-vis* older workers reduces the demand for young labor in mature sectors, although perhaps not in more technologically dynamic parts of the economy. Indeed, one of the more curious aspects of technological change under contemporary circumstances is that new knowledge is embodied in recently educated workers: the young. To the extent that well-schooled young workers bring ideas and techniques that displace older technologies and occupations, young workers may push older workers to the margins of the labor market. But the nature of economic inequality in a knowledge-based economy is likely to limit the effect that embodied technological change has on the employment prospects of workers across the age spectrum, since well-schooled young workers are more likely than not the progeny of highly educated and prosperous adults, who are in turn not only more experienced but also frequently the creators and developers of the technology and knowledge being taught to privileged younger workers.

The foregoing analysis has a very unpleasant implication: in an era of fiscal conservatism, and with structural economic limitations on the capacity of the state to use macroeconomic policy to boost employment and create a "high pressure" economy, young and mature workers become de facto antagonists in labor markets.[1] This line of thinking is discouraging because it suggests that otherwise promising policy approaches that offer the hope of improving the wellbeing of younger workers could well come at the expense of older workers. Policies that reduce the cost of younger workers relative to their mature counter-parts change the composition of unemployment without having much effect on the overall level of unemployment, which is largely determined by aggregate demand. For instance, consider supply-side and demand-side wage subsidies— either in the form of a payment to workers in addition to their market wage (such as the Earned Income Tax Credit [EITC] in the United States, a variant of the Nobel laureate Milton's negative income tax [Friedman 1962]), or as a payment to employers to lower the cost of young workers (as proposed by the Nobel laureate and unorthodox conservative Edmund Phelps of Columbia University). Both are entirely consistent with the most orthodox, free-market economic thinking, and at first sight seem very effective (Phelps 2007). Both have the considerable virtue of being market-friendly, as labor market policies that promote egalitarian ends while improving overall economic efficiency. Yet neither of these ideas stands any chance of improving the overall wellbeing of workers so long as fiscal conservatism puts a damper on aggregate demand.

Let's take a detailed look at Phelps's wage subsidy proposal. The economic and social costs of unemployment, particularly among the young and very poor in the form of crime, as well as the usual and destructive intergenerational transmission of class disabilities from one generation, are logically identical to other sorts of "negative externalities"—the spillover costs of economic transac-tions that damage the wellbeing of third parties. Just as the costs of air pollution due to the use of gasoline-powered cars (lung damage, asthma, and potential climate change) are paid for by people who did not enjoy the benefits associated with producing, selling, or using the gasoline involved, so to do the victims of crime as a result of poverty and youth unemployment bear the brunt of costs associated with the self-interested deals between incumbent workers and employ-ers. Phelps's plea that governments provide a wage subsidy so that employers might hire otherwise unprofitable workers is rock-solid mainstream economics because it suggests that the cost of the subsidy is less than the various costs associated with poverty and youth unemployment. A demand-side wage subsidy boosts the level of job offers for low-productivity workers and young workers by paying employers to hire workers up to the point where the benefits of greater youth employment just balances the social costs of youth unemployment. The positive impact of a demand-side wage subsidy program on employment is independent of whether the labor market is competitive or monopolistic: profit-hungry employers will always hire more workers when labor is cheaper so long as there is a market for their products. In addition, a policy that reduces the cost of labor while at the same time boosting worker incomes will also improve the competitiveness of the national economy in the global system, thereby over-coming one of the few legitimate critiques of minimum wage laws, which, again, make labor more expensive.

While the precise size and structure of a demand-side wage subsidy program comprise a complicated policy matter, the basic idea fits neatly within the corpus of even the most conservative approaches to economic policy precisely because sensible policy seeks to maximize economic efficiency by making sure that prices reflect costs. Indeed, there are many analysts that doubt the efficacy of demand-side wage subsidies due to the administrative costs of an approach.[2] In particular, analysts wonder if a wage subsidy paid to employers of low-wage workers, who tend to be young workers, won't simply be put in firm owners' pockets rather than resulting in pay increases. Anything that increases the profit-ability of hiring low-wage workers will boost the demand for these same work-ers, however, thereby leading to an increase in their desirability and, via competition, a rise in their rate of pay. The one major hurdle to an effective wage subsidy program is, once again, a low level of aggregate demand; subsidizing wages will do nothing to boost employment among young, low-wage workers if there is high unemployment.

Alas, restrictive aggregate demand policies turn a demand-side wage subsidy of the sort Phelps favors into an object of conflict between older and young workers for the reasons noted above. Supply-side wage subsidies aimed at the young do not offer any greater hope than demand-side subsidies, for the same reason: any policy that makes younger workers cheap relative to older workers, but does not expand overall employment, will be a site of generational conflict. Europe's recent and failed experiment with "expansionary austerity" is a case in point. Concern over exploding budget deficits made worse by the Great Reces-sion of 2007–2009 encouraged members of the European Union (EU) to drastically reduce government spending and cut back on a wide range of social benefits on the theory that Europe's long-term economic problems were caused by excessive public-sector debt and excessively high wages relative to the productivity of labor due to a too-generous safety net. The end result of this experiment has been that the working people of Europe, especially the young, have been subjected to completely unnecessary suffering because the presumed disincentive effects of the welfare state on long-term European growth could and can be met by design changes in the system of social supports that reduce labor disincentives without punishing millions of people with poverty. As predicted by many economists, the drive towards austerity across the EU has resulted in high and sustained levels of joblessness because of the collapse of economic activity due to government spending cuts, as well as in an explosion of poverty linked to both unemployment and the removal of social safety nets (Stiglitz 2014). As we write this chapter in October 2017, a popular revolt against high unemployment—including persistent and very high rates of youth unemployment—as well as economic stagnation across the EU has powered the resurgence of right-populist and neofascist political parties across the continent, and it has led to the United Kingdom's choice to leave the EU entirely. The sacrifice of workers, especially young workers, on the altar of fiscal probity is one of the great and needless mistakes of modern economic life.

Squandered youth

Our analysis suggests that there is no necessary conflict between the young and the old in the labor market if economic policy makes room for all workers. This can be done by boosting demand at the same time as the structural impediments to employing the young are reduced, thus making younger workers cheaper by either a demand- or supply-side wage subsidy. Of course, this sort of policy regime requires the state to intervene in economic affairs much more forcefully than the present political consensus in the United States permits. Then again, the contemporary dominance of conservatism in economic affairs is, in fact, a highly interventionist approach tilting towards capital and against labor.

The nature of youth unemployment forces us to choose whether, why, and by how much we are willing to allow luck to matter in the determination of wellbeing. Societies must make hard choices about the balance between the luck of the birth lottery, the luck of the business cycle, the fact that the young are less reliable than the old as workers, and the bias of fiscal policy against the young in democratic market societies. Societies must design social and economic policies smartly lest they make matters worse for young workers, whether by offering rotten schooling, or by making them even more costly *vis-à-vis* old labor and its incentive-incompatible welfare policies, or by allowing the old to exploit the young as shock absorbers in economic downturns. No understanding of economic justice—not even the Hayekian stance that "economic justice" is an illusion (Hayek 1978)—can tolerate a system where the prospects for human wellbeing depend entirely on accidents of birth or avoidable bad luck. Ronald Dworkin (2000) has taught us that economic justice, at a minimum, requires our rules for political and economic cooperation to treat each person's life as equally valuable in the eyes of public power. In a market society, particularly one that relies heavily on the market for labor to distribute economic wellbeing, this basic respect for the real equality of persons to act on their own behalf requires that we arrange our economies so that all children have a very good chance of living as independent persons and surviving the vagaries of commerce with dignity and in fairness. Dworkin (2000, 2011) and Rawls (1971) before him have shown that no society seeking to build a fair economic system could deprive children of the right to develop because their parents are too poor to buy the necessary developmental resources, or allow the old to impose the costs of bad luck on the young.

Appendix

We begin with a standard model of aggregate supply and demand that appears in many textbooks. The supply of output or the level of real national income (Y) in an economy is the result of the use of capital (K) and workers (E) in production according to a neoclassical production function of the form $Y = F(K, E)$ where F satisfies modified Inada conditions; the marginal products of labor and capital are positive and declining (or at least non-increasing), $F_K > 0$, $F_E > 0$, $F_{KK} \leq 0$, $F_{EE} \leq 0$ and $F_{KE} = F_{EK} \geq 0$. The level of employment E is bound from above by the size of the labor force (L), which comprises older workers (L_o) and younger workers (L_y) such that $L = L_o + L_y$. In turn, older workers are much more productive than young workers. We can represent this circumstance by noting that each younger worker is the equivalent of $\Delta < 1$ older worker. Of course, total employment is the sum of employed older workers E_o and equivalent younger workers (E_y) such that $E = E_o + E_y$.

Of course, the actual level of employment depends on the level of aggregate demand as well as on the supplies of capital, labor, and technology available to meet demand. Basic macroeconomics tells us that the level of spending in a closed economy is the sum of consumption spending (C), business investment spending (I), and government spending (G). In turn the primary driver of consumption spending is buying by workers financed by wages. If w_o and w_y are the level of wages paid to older and younger workers, then the level of consumer spending is $C = a(w_o E_o + w_y E_y) = a[w_y E + (w_o - w_y) E_o]$ where $0 < a < 1$ is the average marginal propensity to consume. Investment spending rises with the rate of return to capital, which is tightly connected to the marginal product of capital (F_K) in the standard macroeconomic model taught to most undergraduates. In turn, the rate of return to capital rises as the level of employment increases (this being an implication of one of the Inada conditions, $F_{KE} > 0$, as well as common sense since greater employment means more spending by workers and more profits garnered by firms). If the rate of interest set by financial markets in light of the base rate established by the monetary authorities is r^* then investment can be represented by the function $I = H(F_K - r^*)$, where $H' > 0$ is the derivative of investment with respect to the rate of return to capital net of interest and $I_E = H'F_{KE} > 0$ is the derivative of investment spending with respect to the level of employment. Given the level of government spending (G) the level of aggregate demand is therefore

$$AD = C + I + G = a[w_y E + (w_o - w_y) E_o] + H(F_K - r^*) + G$$

and the level of aggregate employment (E) is determined by the balance between the supply and demand for output $F(K, E) = AD$.

Figure 2.2 illustrates the connections between output market balance and the split of employment between older and younger workers. Output market balance (which is just the very old but still useful idea of the IS curve from basic macroeconomics) in this context determines the position of the IS curve in (E_o, E_y) space, which represents the overall level of employment in the economy. At any point in time the economy can be represented by a point of

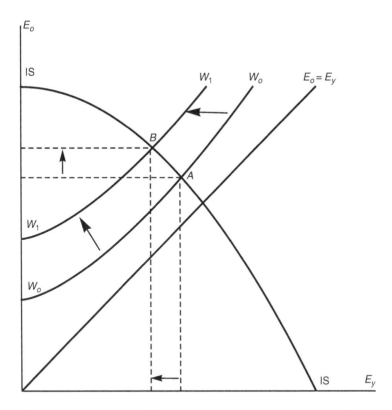

Figure 2.2 Aggregate demand and the age structure of employment (derived from a model developed by the authors)

the IS curve. Also, the higher the level of aggregate economic activity the higher the overall level of employment and therefore the higher the IS curve. Note, though, that the prevailing composition of employment between older and young workers will depend on the hiring decisions of firms based on the cost of each type of worker. The optimal combination of old and young workers is that which minimizes the overall cost of labor. Per basic microeconomics, if the labor markets for older and younger workers can be characterized by the efficiency wage approach as developed by Stiglitz and Shapiro (1984) and refined by Bowles (2004), then the real wages for older and younger workers are increasing functions of their respective employment rates, $w_o = w_o\left(\frac{E_o}{L_o}\right)$, $w'_o > 0$ and $w_y = w_y\left(\frac{E_y}{L_y}\right) = w_y\left(\frac{E-E_o}{L_y}\right)$, $w'_y > 0$, and the optimal combination of older and younger workers is determined by the arbitrage condition $w_y = \Delta w_o$, which in turns implies that the employment rate for younger workers is lower than that for older workers. The optimal combination of older and younger workers chosen by firms is represented by the line $W_o W_o$ in Figure 2.2 and the economy's pattern of employment is represented by the point A. Any development in technology or policy that increases the cost of younger workers relative to older workers will lead to firms substituting older for younger workers, which is shown in Figure 2.2 as $W_o W_o$ moving leftward and perhaps becoming steeper (to $W_1 W_1$), causing the system to move from point A to point B. For instance, a policy that increases the cost of younger workers to employers—say through an increase in minimum wages—will move the economy from point A to point B, thereby causing employers to substitute older for young workers in the short term (and perhaps replacing labor by machines in the face of higher overall wages in the longer term) so long as macroeconomic policies or other economic events do not boost the overall demand for labor as represented by an outward shift in the IS curve.[3] Note that a wage subsidy for younger workers ($\sigma > 0$) changes the wage arbitrage condition to $w_y - \sigma = \Delta w_o$, thereby moving the WW schedule downward and rightward.

Notes

1 We illustrate the tensions between younger and older workers imposed by neoliberal fiscal restraint by exploring a simple model of employment and aggregate demand (in the Appendix) that is implicit in the sort of analyses that all undergraduate economics majors see in their intermediate macroeconomics courses. The simplicity of the model belies a very important point: many potential policy solutions to the problem of youth unemployment that try to reduce the cost of young workers relative to older workers will, in the absence of complementary policies that boost overall labor demand, reduce the employment prospects of older workers.

2 Jaenichen and Stephan (2011) have studied the impact of targeted wage subsidies on the employment prospects of hard-to-place workers in Germany. The subsidies tended to boost the level of employment among the targeted populations, although the overall level of wages did not seem to rise. By contrast, a study of wage subsidies in Finland by Huttunen et al. (2013) indicated that the wage subsidies aimed at older workers had little overall effect on employment but did boost the working hours of the already employed, contrary to the expectations of proponents of wage subsidies. In addition, Brown et al.'s (2011) calibration study into the effects that hiring versus wage subsidies have on wages and employment suggests that the former are far more efficient and effective than the latter in generating long-term job gains for low-ability workers.

3 The slope of the IS curve in $\left(E_o, E_y \right)$ space is

$$\left. \frac{dE_o}{dE_y} \right|_{IS} = -\frac{\Delta F_E - a w_y - a \frac{E_y}{L_y} w'_y - \Delta H' F_{KE}}{F_E - a w_o - a \frac{E_o}{L_o} w'_o - H F_{KE}} < 0$$

and the slope of the EE schedule is

$$\left. \frac{dE_o}{dE_y} \right|_{EE} = \frac{w'_y}{\Delta w'_o} \frac{L_o}{L_y} > 0.$$

Further, the vertical intercept of the IS curve is larger than its horizontal intercept so long as $0 < \Delta < 1$. In addition, the EE schedule is above and to the right of the 45-degree line in Figure 2.2 if $0 < \Delta < 1$, but coincides with the 45-degree line if older and younger workers are equally productive ($\Delta = 1$).

References

Akerlof, George. 1984. *An Economic Theorist's Book of Tales*. Cambridge: Cambridge University Press.

Bowles, Samuel. 2004. *Microeconomics: Behavior, Institutions, and Evolution*. Princeton, NJ: Princeton University Press.

Brown, Alessio, Merkl, Christian, and Snower, Dennis. 2011. "Comparing the Effectiveness of Employment Subsidies." *Labour Economics*, 18(2): 168–179.

Dworkin, Ronald. 2000. *Sovereign Virtue: The Theory and Practice of Equality*. Cambridge, MA: Harvard University Press.

Dworkin, Ronald. 2011. *Justice for Hedgehogs*. Cambridge, MA: Belknap Press.

Friedman, Milton. 1962. *Capitalism and Freedom*. Chicago, IL: University of Chicago Press.

Hayek, F.A. 1978. *Law, Legislation and Liberty, Volume 2: The Mirage of Social Justice*. Chicago, IL: University of Chicago Press.

Huttunen, Kristiina, Pirttila, Jukka, and Roope Uusitalo. 2013. "The Employment Effects of Low-Wage Subsidies." *Journal of Public Economics*, 97: 49–60. Retrieved from https://faculty.smu.edu/millimet/classes/eco7377/papers/huttunen%20et%20al%202013.pdf

Jaenichen, Ursual, and Gesine Stephan. 2011. "The Effectiveness of Targeted Wages Subsidies for Hard-to-Place Workers." *Applied Economics*, 43(10): 1209–1225.

Kahn, Lisa. 2010. "The Long-Term Consequences of Graduating College in a Bad Economy." *Labor Economics*, 17(2): 303–316.

Lindbeck, Assar, and Snower, Dennis. 1989. *The Insider Outsider Theory of Employment and Unemployment*. Cambridge, MA: MIT Press.

Okun, Arthur. 1962. "Potential GDP and Its Measurement." *Cowles Foundation Paper 190*, Yale University. Retrieved from https://milescorak.files.wordpress.com/2016/01/okun-potential-gnp-its-measurement-and-significance-p0190.pdf

Phelps, Edmund. 2007. *Rewarding Work: How to Restore Participation and Self-Support to Free Enterprise* (2nd edn). Cambridge, MA: Harvard University Press.

Rawls, John. 1971. *A Theory of Justice*. Cambridge, MA: Harvard University Press.

Sen, Amartya. 2009. *The Idea of Justice*. Cambridge, MA: The Belknap Press of Harvard University.

Slemrod, Joel, and Bakija, Scott. 2008. *Taxing Ourselves: A Citizen's Guide to the Debate over Taxes* (4th edn). Cambridge, MA: MIT Press.

Stiglitz, Joseph. 2001. "Information and the Change in Paradigm in Economics." Nobel Prize Lecture. Retrieved from www.nobelprize.org/nobel_prizes/economic-sciences/laureates/2001/stiglitz-lecture.pdf

Stiglitz, Joseph. 2014. "Crises, Principles and Policies: With an Application to the Eurozone Crisis." In *Life after Debt: The Origins and Resolutions of Debt Crisis*. J.E. Stiglitz and D. Heymann, eds. Houndmills: Palgrave Macmillan, pp. 43–79.

Stiglitz, Joseph, and Carl Shapiro. 1984. "Equilibrium Unemployment as a Worker Discipline Device." *American Economic Review*, 74(3): 433–444.

3 Precarity in Japan

After the "Lost Decade"

Heidi Gottfried

There is a palpable malaise, both embodied and structural, due to the lingering and intractable economic crisis in Japan. This sensory experience of precarity is deeply felt in everyday life as increasingly more people confront economic hardship and uncertain futures. Youth unemployment alongside large swaths of precarious work expose social fissures and new fault lines of social inequalities. The disaster of the 3.11 earthquake further laid bare Japan's descent from its height as an economic powerhouse. This chapter scrutinizes trends and tendencies towards widespread joblessness among young people, while also examining the challenges posed by nonstandard employment (temporary, part-time, and casual) in the context of an increasingly precarious Japan. Through an analysis of precarity both as effect and affect, the chapter traces the specific trajectory of a worldwide phenomenon within a country once celebrated as a high trust system generating high economic performance. Often cited for its low unemployment rate, both absolutely and relative to other countries, conventional renderings of Japanese capitalism masked the extent of precarious nonstandard work already present in the labor market. Few scholars had noticed that nonstandard employment has become a fixture of the institutional architecture that frames capital, labor, and the state.

By identifying a corporate-centered male-breadwinner reproductive bargain (Gottfried 2013), this chapter sheds light on the institutional logic behind escalating precariousness and its impact on the "lost generation" of young men entering a transformed labor market. Japan's lost generation faces different constraints and opportunities than those faced by their parents (Brinton 2011: 13). Standard employment had provided the material basis of, and cultural conditions sustaining, the hegemonic masculinity of a corporate-centered male-breadwinner system. The recent growth of unemployment and nonstandard employment compels an examination of the changing norm of the male standard work narrative that, until recently, had been so prevalent in Japan.

In this chapter I address in four main sections the puzzle posed by current economic deprivation after decades of rapid economic growth. The first section presents empirical data on employment trends and interprets precarity in the context of changing labor market conditions and the unraveling of the reproductive bargain. To fully understand precarity during the Lost Decade—a period of economic stagnation that began in 1991—the second section looks back at the historical foundation of Japanese capitalism. In brief, Japan's celebrated economic miracle had supported a form of company citizenship, basing rights and entitlements on the masculine embodiment of labor associated with the employment of a core male workforce in large enterprises. This model of capitalism became unsustainable for larger numbers of workers once the economy slowed down during the 1990s. An elaboration on taxonomies of precarious labor categories in the third section illustrates the state's investment in the mythologization and mystification of standard employment. The state's discursive production of precarious labor categories shaped the liminal spaces that nonstandard workers occupy. Such liminality and precarity have prompted the emergence of new subjectivities and organizational practices around gender and class relations in flux. The fourth and final section focuses on the formation of new labor associations organizing the unemployed and precarious workers in a changing landscape of work and politics.[1]

Interpreting youth unemployment and precarity

Prior to the bursting of the economic bubble in 1990, younger workers aged 15–24 exhibited low unemployment rates: only 4.5% for men and 4.1% for women (see Table 3.1). Since the onset of the Lost Decade, the unemployment rate doubled for both male and female younger workers, reaching 9.4% for the youngest workers aged 15–19, 9.1% among those aged 20–24, and declining to 7.1% for those aged 25–29, as compared to 5.1% overall in 2010 (Japan Institute for Labour Policy and Training [JILPT] 2012: 24). Five years later, youth unemployment had dropped to 5.5%, which was still above the pre-bubble level (JILPT 2015: 221–223). From a cross-national comparative perspective, Japanese young people seem to fare better in the labor market. Japanese unemployment statistics, however, fail to fully represent the extent of unemployment, only counting as unemployed those individuals aged 15 and over who are out of

Table 3.1 Unemployment by age and gender (1990–2010)

	1990	*2000*	*2010*
Total Overall	2.1	4.7	5.1
Male Total	2.0	4.9	5.4
Male 15–24	4.5	10.2	10.4
Male 25–34	1.8	5.0	6.6
Female Total	2.2	4.5	4.6
Female 15–24	4.1	7.9	8.0
Female 25–34	3.4	6.4	5.7

Source: JILPT (2013: 42).

work, "capable of immediately accepting work," and seeking work during the survey period (JILPT 2013: 44). As in the United States, this definition of unemployment omits the non-employed, including discouraged job seekers opting out of the labor market, whether by choice or by circumstance. Furthermore, the dichotomous framing of employment and unemployment in conventional statistical language obscures employment statuses in-between these labor categories, and therefore underestimates the true levels of less than full employment and the extent of nonstandard precarious work.

In Japan, the definition of nonstandard employment refers to employment statuses deviating from standard full-time, continuous employment and its associated statutory and workplace benefits.[2] The main forms of nonstandard employment include part-time, temporary (agency and direct-hire), casual day labor, and independent contractors. Overall, the proportion of 15–19-year-old workers in nonstandard employment jumped from 20% in 1982 to 75% in 2007. During that same timespan, nonstandard employment among those aged 20–24 nearly quadrupled from 10% to 40% (Brinton 2011: 28).[3] For male workers in the 15–24-year-old range, nonstandard employment grew from 8% in 1991 to just below 30% by 2013. Interestingly, young women's nonstandard employment at 10%, only a few percentage points more than men in 1991, mushroomed to 40% 22 years later (JILPT 2014: 53). A special category called *fureta*, signifying young freelance workers, totaled 490,000 men and just over 1 million women in 1997. The number of male *furetas* doubled to 980,000, while the number of female *furetas* increased to 1.9 million by 2003; these numbers slightly declined for men to 920,000 and majorly reduced for women to 950,000 as of 2007 (Cook 2013: 30). Further, 2013 data tabulated from the Ministry of Internal Affairs and Communications' special Employment Status Survey found, perhaps unsurprisingly, that educational attainment correlated with employment status for 25–29-year-old men. Male high-school graduates were less likely to secure permanent employment (not changing jobs after leaving school) than university graduates: 39.4% compared to 57%, respectively. Correspondingly, men with only a high-school degree exhibited higher continuous nonstandard employment at 12% and more than double the chance of moving from regular to nonstandard employment (5.2%) than male college graduates (JILPT 2014: 54).[4] Taken together, the picture that emerges is one of extensive underemployment and precarity.

Precarity registers not only in the churning of workers into and out of the labor market, but also links to the ways in which we work (Kalleberg and Hewison 2013) and inhabits our senses (Allison 2014). Various theorists have analyzed "precarity" and precariousness in terms of risk, as an ontological condition of human insecurity (Butler 2006), and as a descriptor of uncertain work/life conditions emblematic of liquid modernity (Allison 2014; Lorey 2011). But precarity, precariousness, and precarious work map different dimensions of social, cultural, and economic phenomena. While precariousness may be a property of being human, precarity encompasses a widespread "social form" (Chang 2011), reflecting increasingly market-mediated conditions of uncertainty and insecurity, materially and existentially (Arnold and Bongiovi 2013), both as effect and affect. In the first instance, precarity designates the effects of different political, social, and legal modes associated with neoliberal governance and post-Fordist employment (Lorey 2011; Rosenbaum 2014: 5). Also, in terms of effect, precarity is associated with temporal and spatial irregularities of work that disqualify workers from social protections. Such precarious work refers to employment lacking both implicit (derived from past practices) and explicit (arising out of contractual rights) guarantees for long-term employment (Kalleberg 2011). By affect, Grossberg (2013: 461) "describes complex articulations among imagination, bodies, and expressions." As "embodied affect," Allison (2015) describes how precarity "registers on the senses in the first place—as a sense of being out of place, out of sorts, disconnected." In this latter sense, precarity is a disposition of self *vis-à-vis* society, following Butler, who reserves precarity for identifying that "class" relegated to the margins by the state (as cited in Allison 2014: 64–65). At a societal level, precarity juxtaposes the semblance of "ontological security" that had characterized the traditional social order with the "turbulence of the risk society [...] that is not open to calculation by individuals or by politics" (Beck 1999: 11–12).

The growth of precarious work and the expansive sweep of precarity are each a symptom and a result of a redistribution of risks from state and economy to individual workers and families. Economic insecurity and precarity preceded the crisis of the 1990s; they were built into the Japanese variety of capitalism. A brief history of the post-war consolidation of Japanese capitalism highlights precariousness in its early manifestations and the specificity of precarity in its current forms.

Japanese capitalism and precariousness

The development of Japanese capitalism, held up as a model for economic prosperity and growth, underplayed the role of nonstandard labor and precarious work in the narrative of success. Historicizing the Japanese form of capitalism and its reproductive bargain can reveal precarious work going largely unnoticed during the economic expansion of the 1960s and 1970s. This model's seeming strength turned into weakness for workers and their unions tied to the fate of the firm, and for the increasing number of workers left out of the bargain when the Japanese economy entered the Lost Decade of the 1990s.

Japanese capitalism "operated through a broad configuration of disciplinary institutions, hegemonic rule through creation of social consensus and normativity, and forcing of individual and collective identities in complex relation to one another" (Yoda 2006: 35). Salarymen were at the center of this model of capitalism and the male-breadwinner reproductive bargain. While the term *salaryman* harkens back to the Taisho period (1912–1926), the expression of heteronormativity signified by this hegemonic construct gained currency in circulation during the 1960s (Roberson and Suzuki 2003: 7). The ideology of the salaryman or the male breadwinner both allowed for and compelled men to devote long hours as workers in capitalist corporations and to fulfill roles as taxpayers (Roberson and Suzuki 2003), as juxtaposed against women who were expected to occupy positions as "good wives and wise mothers" on behalf of the nation (Uno 1993). Many of the themes embodied in the image of masculinity also constituted the core composition of adulthood; that is, taking responsibility towards society, family, and work organization (Cook 2013: 33). The large numbers of working-class women who worked full-time in smaller manufacturing firms—many in the textile industries before the decline of the sector during the 1980s—were subjects of and subjected to the ideology of the male breadwinner.[5] The Japanese welfare system based entitlements on this male-breadwinner model so that women typically derived rights and benefits as wives and mothers rather than as workers (Gottfried 2009).[6] This configuration of class, gender, sexuality, and nationality informed the inscription of the ideal family and worker, which participated in the structuring of hierarchies and exclusions (Bergeron and Puri 2012: 497).

In Japan, the social contract and regulatory regime anchored social protection in large corporate structures in support of a strong male breadwinner. This corporate-centered reproductive bargain established a strong form of "company citizenship" coordinating worker's interest with the economic prosperity of their firm. The resultant weak labor organization at the national level, and high fusion of labor representation with employers at the enterprise level, deprived labor of a strong political lever for realizing a social bargain. From this position of relative weakness, unions settled for strong internal labor markets that ensured job security, in-house training, an age-graded system of rewards, and an array of benefits (e.g., end-of-year bonuses, housing and transportation allowances, and pensions) for their members in large industrial companies. Through the *nenko* seniority system, management created a path along which men could rise to supervisory positions. Kumazawa (1996: 167) argues that the removal of women from competitive career tracks enabled management to open a secure route for male workers. This practice contributed to the fusion of labor representation with company organization interpolating workers as male "company citizens" (Gordon 1985). The community-type institutions that fostered company citizenship rarely extended beyond the core male workforce (Tabata 1998: 199).

The practical and conceptual basis of company citizenship was built on the assumption of an exclusionary bounded political community, creating a stark insider/outsider divide with its gender inflections. Job security of regular employees working in large Japanese companies was underwritten by the less visible inferior working conditions further down the job hierarchy and the production chain throughout the post-war period (Gill 2003). The economic expansion that followed during the 1960s stabilized employment relationships for many men who moved into more secure industrial employment in unionized enterprises, while these same enterprises and their subcontractors employed men in nonstandard employment as a flexible labor force (Slater 2009). Women, whose jobs were concentrated in small- to medium-sized firms and in nonstandard employment, fell outside union coverage, lowering their access to social wages and the benefits of worker citizenship determined at the enterprise level. Moreover, ideological and economic pressures pushed many women out of the labor force. A large percentage of women left the labor market after marriage and/or childbirth and reentered when their children reached school age—showing up statistically in the M-curve employment diagrams.[7]

Since the mid-1970s, when shocks from the global energy crisis sent tremors throughout this oil-dependent nation, the rate of growth and an increase in the absolute numbers of nonstandard employees highlighted a more noticeable staffing strategy. Japan's flexible mass production model further incorporated nonstandard employment, chiefly among women, as an untapped, cheap labor buffer to manage high personnel costs associated with the lifetime employment system and with the male-breadwinner reproductive bargain. The economic downturn of the 1990s, due both to stagnation and to increasing global competition, further eroded the male standard in favor of other employment statuses. Firms subsidized high wages and the security of its aging and shrinking core workforce by expanding non-standard employment. While special work visas offered to overseas workers of Japanese descent opened a side door for unskilled workers to take up employment in low-level jobs in manufacturing, overall immigration policies foreclosed the possibility of employers filling the large and growing demand for nonstandard employment primarily with low-wage migrant labor.

Although the promise of lifetime employment has not disappeared as either a practice or an expectation, regular jobs are converted into, and new jobs are created in, forms of nonstandard employment, most prominently for women, young people, and older workers. But also salarymen have lost the security once provided by the "three treasures" of lifetime employment, seniority-based pro-motion, and company unions (Roberson and Suzuki 2003: 9). Employment insecurity undermines the ideological and economic basis of the old hegemonic reproductive bargain represented by the salaryman.

Consequently, the growing number of young men employed in precarious work challenges the dominant image of Japan represented symbolically by the heteronormative, middle-class, citizen worker in Japanese culture and politics (Dasgupta 2003, 2012). New entrants and workers reentering the labor market often end up in nonstandard employment, further destabilizing career paths over the life course. It is now well documented that young men and women no longer can expect to find a permanent job right out of school. Brinton (2011) chronicles the fate of the current generation of young non-elite men "lost in transition." The institutions for skills development acquired on the job through the training system over one's working life now create instability for young people in post-industrial Japan. Faced with instability, many are postponing family formation and fertility decisions. Persistent nonstandard employment is rooted in the design of the institutional architecture supporting pillars of the Japanese employment system, the reproductive bargain, and its mode of state regulation.

The language of precarity and precarious labor: creating the lost generation?

Standard employment is the dominant frame of reference determining those subjects worthy of social protection and legal recognition of citizenship rights and entitlements. In Japan, the specific definition of standard employment refers to "those employed immediately after graduating school or university and have been working for the same enterprise" (Ministry of Health, Labour and Welfare 2006). This official government definition assumes the institutional mooring of standard employment in the tightly coordinated relationship between education and employment, facilitating the transition from school to a port of entry into a firm-specific internal labor market with the expectation of long-term continuous employment as part of the lifetime employment system, a practice rather than a principle written into contracts (Gottfried 2015). Generous benefits associated with standard employment are rooted in company citizenship. By contrast, the state provides minimal statutory social protections and entitlements, and indexes benefits to the standard employment relationship.

Conventional labor statistics and labor regulations use this historical standard as the master category for framing a hierarchy of employment statuses. This dichotomy leaves out non-waged work and employment statuses in-between these categories. Non-employment does not simply demarcate non-work or the absence of employment, but also glosses over non-market exchanges, ranging from unpaid reproductive labor to cooperative exchanges outside the orbit of traditional employment relations. In conventional formulations, only wage labor deserves full recognition of rights, rewards, and social protections. Labor statistics, conventions, and regulations privilege and naturalize wage labor as the primary means of gaining political recognition and economic security.

Still, precariousness is not only an outcome of less regulation of nonstandard employment and deregulation, but also a consequence of differential rights, risks, and rewards inscribed in the language used to frame precarious labor categories. By inventing new terms and re-signifying keywords, the state discursively shaped the gendered and age-graded profiles of precarious work and workers.[8] Somewhat uniquely, the Japanese use loan words to indicate that the concept is of "foreign" (usually Western) origin in a kind of "inverse Orientalism" (Hirakawa 1995: 94), which differentiates these employment forms from the standard equated with Japan and sets them apart from Japan, "distancing itself from the Other that is the West" (Endo 2006: 1). Deploying loan words rhetorically served to detach the employment form from the exploitative structures of capitalism in Japan. A review of linguistic shifts in the national classification of employment statuses can be read as expressions of "instrumental practices in regimes of power" (Krishnan 2014) through which "a state produces segmentation of the citizen body" (Fourcade 2010: 572).

The frame of reference for determining part-time employment would seem rather straightforward based on an agreed-upon conventional threshold of hours worked below a full-time standard. In the early 1990s, Japanese labor law enshrined a legally permissible distinction defining part-time employment relative to full-time hours at an enterprise rather than the usual economy-wide standard number of hours worked. The Part-Time Workers' Law of 1993 defined "short-time workers" as those working shorter hours than regular workers employed in the same "undertaking" (Kurokawa 1995: 57). In some government statistical series, a short time referred to those who worked less than 35 hours in the week the survey was conducted, including seasonal workers and irregular workers. Part-time (*paato-taimu*) was most commonly defined in negative terms, encompassing any position that was not a regular position (*sei-shain*), a position that involved shorter hours than *sei-shain*, and one that employers merely designated as such. *Sei-shain*, in turn, referred to an employment relationship of an unspecified duration. But like *paato-taimu, sei-shain* was a category left largely to the discretion of a firm's human resources office. As a result, some workers classified as part-time actually worked 40 hours per week. The inferior employment status is a function of legal classification rather than a threshold of hours worked.

The state also coined keywords for the purpose of establishing differential legal statuses by inventing a new legal nomenclature, a tabula rasa for signifying agency temporary work. To avoid a specific historical referent, temporary-help firms and legislators minted the Japanese word *haken*, which translates to "dispatch," as a more neutral label that would disassociate agency temporary employment from the negative connotations of temporary employment in Japan,[9] in an effort to overcome the stigma of unskilled labor associated with agency temporary employment in the United States. The memory of the exploitation of temporary workers during the 1950s informed the framing of the Dispatching Law (1985) that explicitly prohibited the use of temporary employment in some traditional manufacturing jobs.[10]

The terminology of *haken* specifies the triangular employment relationship between a labor market intermediary, a client firm, and an employee. As elsewhere, this triangular relationship adds a third party between the employer and the employee. Although labor law holds the temporary-help firm responsible for supervising and establishing the work conditions of the temporary employee, the reality of the triangular relationship creates legal ambiguity for temporary employees seeking to exercise their rights both at the client companies where they work and *vis-à-vis* the temporary-help firms that employ them. More specifically in Japan, legal ambiguity is in part the result of labor laws and legal norms based on a standard masculine work narrative rooted in the corporate-centered male-breadwinner model. The exclusion from a firm's internal labor market deprives *haken* of job security and those benefits conferred by the company bargain. The legal terminology of *haken* codifies differential rights both *to* and *at* work, which makes this employment form precarious.

Nonstandard employment, such as *paato* for part-time and *arubaito* for a young part-time worker, borrow terms from English and German, respectively. Both *paato* and *arubaito* index the life course in common usage and in labor law. In popular usage, *paato* sometimes combines with *obasan* (middle-aged women), or simply with *-san*, an honorific ending that can indicate Mr./Mrs. This rhetorical linkage of *paato* to *no obasan* has come to imply married, middle-aged women who make up the vast majority of part-time workers. In this sense, *paato no obasan* suggests that the worker occupies only a part of the status of a full-time employee but never a substantial part. Implicitly, the middle-aged woman is seen as a secondary earner who relies on a male breadwinner for income security. On the other side of the age spectrum, *arubaito* designates a young person's entrance into the labor market, referring to a transitional phase for young workers who might take up temporary part-time employment during their course of studies. When initially introduced, *arubaito* denoted a rite of passage in a typical male work narrative before the student entered the "lifetime" employment system. An *arubaito* status was neither expected to last more than a short time nor to negatively impact on long-term, permanent employment in a company.

As the economic malaise of the Lost Decade lingered into the 21st century, and as more men as well as women remained unemployed or in nonstandard employment long after their student hiatus, the state promulgated new keywords that shifted the registers for the interpretation of a range of employment relationships. A proliferation of popular and legal terms chastised young people, as a lost generation, for shirking their work responsibility, thereby strengthening pejorative connotations associated with nonstandard employment. These terms include: *fureta*, for young people, aged 15–34, who are neither students nor housewives and who job hop (Cook 2013: 29); non-employment, including NEET, referring to young people not in education, employment, or training; *parasaito shinguru*, for young adults who are likened to parasites living off the financial largess of their parents; and net-*café* refugees, for the homeless, primarily male young people, whose main domicile is the internet *café* (O'Day 2012). *Arubaito* morphed into the category of *fureta*, combining the English word "free" with *arubaito*. This hybridized linguistic form implied that *free-arbiters* or *furetas* willingly eschewed conventional employment and "freely" chose this nonstandard employment form. Initially, *furetas* projected a positive image of a new entrepreneurial, reflexive, self-directed man; a freelancer who showed initiative in the new economy. Around the same time, the state began using NEET, borrowing from the British expression for young people not in education, employment, or training. The Japanese Ministry of Health, Labour and Welfare, however, expanded on that definition to cover all individuals from the ages of 15 to 34 who were not enrolled in school, not in the labor force (thus, not counted as unemployed since they were not currently looking for work), and not married (JILPT 2014: 54).

Into the new millennium, those in nonstandard employment took on the negative mantle of a lost generation, of being "lost in transition" (Brinton 2011). Young men were represented as slackers, either refusing to become adults or forsaking the career paths of their fathers, and altogether opting out of the old bargain symbolized by the salaryman. Re-signification of keywords projected an image of young men as socially adrift, unproductive members of society, failing to fulfill their masculine citizenship duty as "rights-earning individuals not as needy family members" (Haney and March 2005: 464). Young women in the same employment categories were portrayed as frivolous girls. These categories used an individual moral/psychological register equating nonstandard employment with selfishness and immaturity.

Negative gender connotations also were reflected and inflected in policies critical of *furetas*, who were portrayed as part-time job-hoppers forestalling marriage and childbirth (JILPT 2013: 36). The Work-Life Charter (2007) was one of many legislative attempts to shore up the declining fertility rate (1.25 at the time) by targeting a reduction in the number of *furetas*. Missing from the representation was the structural circumstances leading to the growth of this labor market segment. *Furetas* who do not or cannot pursue a regular career are at a strong disadvantage in the Japanese labor market, where length of service is a proxy for nontransferable skills built while working for the same firm (Boyer 1998: 158), and where standard employment is a basis for claiming many citizenship entitlements.

These keywords both classify and codify a variety of labor relations and nonstandard employment arrangements. It is important to analyze national classification schemes because words/discourses do not merely reflect power relations, but are "active forces shaping" material lives (Fraser and Gordon 1994: 310). Keywords gain their currency in reference to institutional norms, rules, and regulations. In Japan, labor law and employment regulation still derive from labor standards based on heteronormativity inscribing the male work narrative without interruptions for care responsibility. The discursive production of precarious labor categories both describes and assigns work positions. As a result, workers, both male and female, who deviate from this male-breadwinner standard, suffer penalties in terms of foregone promotions and training, lost earnings, and limited pensions; risk social exclusion; and experience dislocation. Through the creation of new employment statuses with inferior work conditions, the state actively produces precarious work and precarity.

Precarity: challenges and opportunities

Precarity presents both challenges to and opportunities for resistance, and for forging new political subjects poised against and formed by liquid modernity. Liquid lives, characteristic of precarious work and working, detach workers from a stable worksite from which to ground union organization and membership. Occupying liminal spaces also ushers in new identities and bases of solidarity. A scan of recent labor organization reveals not only new groups being represented, but also locates new forms of community unions coalescing in diverse places and spaces, and at different scales, each operating alongside traditional unions and locating in the spaces in-between work and civil society.[11] These new unions represent unemployed and nonstandard workers, and organize according to employment status and social location, including, for example, the Precariat, a community union representing workers in nonstandard employment; the Women's Union Tokyo (WUT), a community union targeting women workers in any occupation and industry located in Tokyo; and Tokyo Youth Union of Contingent Workers (Tokyo Youth Union). Changing employment conditions and the unraveling of the reproductive bargain spur innovative organizational responses among workers in precarious employment and among the unemployed.

Precarity and precarious work have long been features of women's wage labor in Japan. Both in law and in practice, the masculine embodiment of the labor subject symbolically annihilated women as workers. Despite (and because of) their exclusion from enterprise unions, feminist labor activists established separate worker associations for representing class and gender-related issues (sexual harassment, pay equity, social reproduction, etc.) neglected by traditional unions. Formed in 1985, the WUT was the first association extending organizational membership exclusively for a female constituency regardless of their employment status.

Following the example of community unions sprouting up during the Lost Decade, and influenced by WUT, the Tokyo Youth Union was formed in 2000 (Gottfried 2014a). The Tokyo Youth Union is the oldest and largest union specifically aimed at mobilizing young workers in unstable jobs and the unemployed. Smaller youth unions populate other areas surrounding Tokyo (Chiba, Kanagawa, and Saitima) as well as in Sendai, Nigata, Nagoya, Okayama, and Okinawa. All of these youth unions are loosely and informally affiliated; some are dormant and inactive, and others have strong ties with the Tokyo Youth Union. The Tokyo Youth Union started with a membership of only 30 individuals, growing to around 350 over the next 10 years, fluctuating slightly from year to year. Members range in age from their early 20s to their early 30s. Many of the original members' serial insecure employment history keeps them involved in union activities. These members, largely single men, reside in cheaper apartments either on the outskirts or in older buildings around Tokyo. Increasingly, members share a small apartment; fewer live with their parents (which reflects the older age profile of their youth membership).

With members living and working scattered across the sprawling city, the Tokyo Youth Union improvises organizational strategies focusing on broad economic themes and legal issues. Individual workers, using hotlines set up by the union, call the number for consultation over labor violations.[12] An unusual Japanese labor law provision allows unions to "collective bargain" for a single worker. Leveraging this legal quirk, the union initiates collective bargaining for the individual employee when notified about a violation of the Labor Standards Act (LSA, 1947). According to the union, companies tend to recognize the standing of the union and proceed to negotiate. In those cases where the company denies the claim and/or refuses to recognize the union as a bargaining partner, the union innovates ways of bringing the company to the bargaining table. For example, it may lodge a complaint at the local labor standard's office (organizationally under the Ministry of Health, Labour and Welfare). Other cases result in lawsuits. A court case brought against a ramen noodle company resulted in a victory when the court found that the company had violated labor law and directed it to make restitution. Collective bargaining of this type involves only narrow legal issues, since the union only "represents" one or a few members— although settlements apply to all affected workers extending the benefit beyond the small number of official members. Thus their success at the bargaining table extends beyond the claimants. Finally, symbolic protests raise the issue in the public eye. Most spectacularly, 500 temporary workers erected a tent village centrally located in Tokyo's Hibiya Park from where they could protest their precarious existence, which was made worse after the collapse of Lehman Brothers in 2008. The Tokyo Youth Union mobilized temporary workers for this spectacle. In so doing, it drew on local idioms of day laborers' makeshift domiciles glimpsed in other city parks for this symbolic protest. The protest raised public awareness of the heretofore invisible precarious work and living conditions of these young workers.

Innovation exemplifies the union's modalities, operating both in civil society and in workspaces. Taking a broad-based approach, the Tokyo Youth Union reaches out around specific campaigns, such as lobbying against the deregulation of the dispatching law and working with non-governmental organizations (NGOs) in their anti-poverty campaign. As a community-based union, its mission extends beyond workplace-based issues to broader social welfare considerations. The precarious existence of these workers led the union to form an anti-poverty network. Although not explicitly recognized as such, the union attends to workers' experience of precarity both as effect and affect.

Despite victories at the bargaining table, the union has not converted its successes into new membership. While the organization is "socially recognized," organizing remains difficult among this population. The union has found that younger workers are unaware of their rights, have little knowledge about unions, and do not realize that their temporary employment is not necessarily temporary. Generationally, young people experience economic dislocation in the context of union decline and quiescence—the early 1970s marked the last major strike wave in Japan. Smaller activist unions share a common financial challenge; their small and poor membership cannot adequately fund union activities, including the hiring of organizers. Unlike the enterprise-based unions, which can rely on dues from a relatively stable membership,[13] these unions actively seek new members across many firms, occupations, and industries to secure resources. Union activists must survive on minimal support from union dues and engage in union activities after work.

The changing nature of work and social precarity may herald the formation of a precariat based on both the loss of long-term employment commitments at the root of organizational career paths, and the disconnection from old forms of social protection and sociality (Standing 2011). Precarious employment as effect deprives people of the ability to establish a long-term career, a vocation, and a livelihood; and precarity as affect is a corrosive condition eroding the foundations for sustaining an identity. As indicated above, precarity has a potentially empowering effect, as nonstandard workers challenge the way we work and live in a capitalist society. New unions populate the political landscape in Japan, but differing visions and structures hamper inter-organizational cooperation that could build social movement unionism. Union activists express ambivalence about their relationship with Rengo, the national labor federation (and less so with Zenroren, the smaller, communist-based labor federation). Although Rengo rhetorically supports organizing the unorganized, particularly precarious workers, its successes have been limited. For example, the WUT has resisted overtures to merge with traditional unions and other community unions. Its need for "a room of one's own" is born out of its feminist commitment that only women can empower women. The Tokyo Youth Union's independence originates in class politics. Ambivalence has structural roots as well. These unions organizing precarious workers do not have a home to call their own from which to ground organization and identity. It is also the case that precarious workers' disparate work conditions militate against labor organizing. These workers are spatially disconnected from each other across dispersed work sites; their temporally uncertain work schedules—working on call or as needed—diminishes reliable channels of communicative action; and their contractually unpredictable employment prevents them from building up social capital. For this reason, organizing precarious workers cannot follow organizational scripts of the past centered on a stable collectivity at the enterprise.

Conclusion: accounting for and the consequences of precarity in Japan

This chapter uncovers vulnerabilities that already existed as part of the economic model but are now becoming more acute. Rather than being merely a residual dimension of Japanese employment practices and structures, the analysis presented here indicates that nonstandard employment represents a key component of work transformation and underscores the salience of class, gender, sexuality, and nationality in the process of Japan's ongoing restructuring. The unraveling of former employment guarantees (i.e., lifetime employment and corporate-based welfare) and the role of the state in framing precarious labor categories create new risks; risks that emerge as more workers are exposed to economic uncertainties and employment insecurities. Generally, nonstandard workers must improvise a sense of belonging amid institutional uncertainty. In addition, temporary workers represent the quintessential "neoliberal selves" who must navigate the social, cultural, and cognitive uncertainties of being socially excluded and disconnected from a fixed place of work and work group. Consequentially, precarious employment leaves workers uncertain of future time horizons.

One of the so-called pillars of the Japanese economic model, the lifetime employment system, was maintained by large firms that guaranteed job security to core male workers; it relied on hidden precarious and unpaid labor, mainly among women. But lifetime employment has been almost an exclusively male domain supported by the unacknowledged corporate-centered male-breadwinner reproductive bargain. In this institutional context, some Japanese men experience unemployment as a social disgrace and consider it shameful to register at public employment agencies. Japanese women may not report unemployment for different social and economic reasons. With few portals into internal labor markets, Japanese women who either lose or cannot find jobs may stop looking for work when confronted by the difficulty of securing reemployment in times of economic crises. These unemployed, discouraged women disappear into the shadow economy of the household as non-employed, no longer counted in unemployment statistics, and expected to derive rights and income security as dependents on a male breadwinner. Thus, the unemployment rate is likely to understate the actual number of men and women out of work.

There are possible far-reaching effects of the rise and persistence of unemployment and nonstandard employment on the lives and livelihoods of young men and women. Fewer workers enjoy access to the social protections tied to occupational and employment status that were the basis of securing consensus at the center of the old reproductive bargain. Particularly with the growing numbers of men in nonstandard employment, we need to consider the consequences of masculinization of precarious employment and precarity on gender identities and power relations, and on new social imaginaries. What will be the effect on core gender and sexual identities as more men at various stages of their working lives occupy these "feminized" positions? How will masculinity be reconstructed when men no longer can earn a family wage or fulfill the obligations of this breadwinner role? One small qualitative study found that male *furetas* desired to create alternative lifestyles, but their narratives about the future still were framed in terms of normative ideas of the productive male citizen. In this sample, male high-school graduates whose fathers worked in blue-collar jobs preferred securing stable employment, while university graduates favored freedom from white-collar work and looked towards carving out their own lifestyle not dictated by the enterprise (Cook 2013: 40).

As we know, the effects of precarity can lead to despair with tragic results, such as the increasing incidence of suicide among young men (Allison 2015). Changing economic circumstances undermine the social and cultural conditions felt acutely by young men who are being denied the normal entry into the labor market. Displaced and disconnected from long-term employment prospects, young men are often unable to cope with social-psychological trauma. Deconstructing media coverage of a murderous rampage by a young man in Akihabara—the electronics shopping district and home to Japanese gaming culture—Slater and Galbraith (2011) parse out discursive evidence of cultural and social contradictions associated with the loss of stable employment among young men in recessionary Japan. In this spectacular case, the male young man's precarious work status was central to the narrative framing the event. What the case demonstrates is the new cultural significance of precarity as embodying loss: loss not only of stable employment, but also of social moorings.[14] Unmoored from the corporate-centered salaryman promise of stable employment, this precarious status in their liquid lives deprives such young men of the social foundations on which male middle-class respectability previously had been based. Ironically, national rhetoric now chastises young men for selfishness and not being selfless contributing members of society, blaming them for structural transformations beyond their control. Young men are being held accountable for not taking responsibility as productive male citizens; their precarity is attributed to personal failings and not to institutional failings.

Japan is entering its third "Lost Decade." One can reflect on the picture emerging from Anne Allison's post-tsunami elegy to Japan:

> The sea of mud that pummeled what had been solid on the coastline signaled something else: a liquidization in socio-economic relations that started in the mid-1990s (but actually before) with the turn to flexible employment and its transformation of work and the workplace.
>
> (Allison 2014: 7)

The tsunami and nuclear disaster at the Fukushima Daiichi plant shook up society, revealing the vulnerabilities that lay dormant in society. Homeless and unemployed men, society's cast-offs, were hired as temporary workers to clean up the dangerous irradiated site (Gottfried 2014b). Three "lost" decades have taken their toll, leaving nothing untouched in their wake. Turning inward is a symptom and a response to uncertain times. Hopelessness and despair are not the only responses to precarity: young men and women in nonstandard employment are organizing into new unions. What impact these fledgling unions will have on policy and politics remains unclear.

The case of Japan opens a window from which to observe the inherent crisis tendencies and increasing risks of neoliberal global capitalism. Such developments in Japan foreshadow risks posed by an old reproductive bargain designed around a set of heteronormative employment relations and economic conditions no longer applicable for a growing number of workers worldwide. Nonstandard precarious work, hidden in the representation of the Japanese economic miracle, now reshapes employment relations and social reproduction in fundamental ways.

Notes

1 This chapter benefited from the participants who attended the workshop on youth unemployment at Middlebury College, braving one of the coldest winters on record. I have incorporated many of the suggestions from the editorial team, Sujata Moorti, Jamie McCallum, and Tamar Mayer, who undoubtedly improved the analysis. Their comments, along with the additional ideas from David Fasenfest, are reflected in the pages of this text.

2 While the nomenclature of standard/nonstandard has applicability to Japan, this binary is problematic as a signifier of precarity for several reasons, including: labor standards vary even among workers in the same categories and across countries; the standard employment relationship has been limited to a small number of workers, primarily men in the unionized manufacturing sector, and differs in countries with strong labor regulation and unions; the standard employment relationship never existed in the global south or among women, but rather was a function of a particular class and gender bargain in the West; and not all nonstandard work is precarious (Bernhardt 2014).

3 In 2007, Brinton speculates that around 11–12% of individuals in their early 20s to approximately 20% of individuals in their late 20s and early 30s were considered non-employed, neither in employment, in education, nor in training (NEET).

4 The government from which these data were gleaned did not report comparable statistics relating education and nonstandard employment among women.

5 A personal conversation with Professor Kimiko Kimoto of Hitotsubashi University reminded me of gaps in the narrative of the Japanese economic miracle. Japan's economy is notable for the relatively large share of women working full-time in jobs in the manufacturing sector, particularly in the textile industry. Historically, the textile industry was a key sector propelling the economic miracle until its decline in the face of global competition from low-wage labor in other parts of Asia. Since the 1980s and 1990s, the downsizing of the textile industry has left many women out of work and underemployed.

6 The male-breadwinner model privileges married women with children over single mothers in the distribution of welfare benefits. Unable to rely on a safety net and unable to access favorable tax deductions, single mothers must work to secure their family's livelihood. Single mothers are the category of women with the highest rates of full-time employment, and the highest rates of poverty.

7 The M-curve characteristic of women's labor force participation was less pronounced in 2011 than in 1975. Most dramatically, women in the age group of 25–29 experienced their peak labor force participation rate of 77.2% in 2011, up from a low of 42.6% in 1975. During the main childbearing years, women's labor force participation fell to 67%, increasing again for many women who returned to work (JILPT 2013: 17).

8 See Imai (2011) and Gottfried (2009) for a comprehensive review of relevant labor and gender regulations.

9 Legal frames circulate through various mechanisms. In some respects, China has emulated Japan in its enactment of weak labor regulations, including the use of specialized language such as "dispatching" instead of "temporary" to identify those employed by agency temporary help firms (Shire 2012).

10 Before the Dispatching Law passed in 1985, temporary employment, particularly agency temporary work, flourished largely unregulated, except for a law prohibiting the operation of private placement firms. Employing temporary workers to gain flexibility was widespread throughout the 1930s and 1950s, decades before the economic miracle took off. At the Hiroshima factory of Nihon Steel Works, three-quarters of the labor force held temporary positions in the 1930s (Hazama 1997: 170). These workers suffered deprivations and coercive treatment on the job. Abuse of temporary workers' rights in the manufacturing sector also was rampant during the 1950s.

11 The discussion of new unions is based on interviews with union representatives conducted over a 14-year time span, including interviews with Tokyo Youth Union representatives on July 12, 2014.

12 Japanese activists set up hotlines for enabling individuals to report on rights' violations.

13 Japanese enterprise-based unions do not have to actively organize new members who join as a function of their lifetime job status. Their relatively stable membership ensures a flow of fees and a conduit for new leaders who rise through the ranks (see Gottfried 2015).

14 Young people's social disenfranchisement and impoverishment are themes taken up in a new "precariat" literary genre in Japan (Karin Amamiya, cited in Rosenbaum 2014: 4).

References

Allison, Anne. 2014. *Precarious Japan*. Durham, NC: Duke University Press.

Allison, Anne. 2015. "Social Precarity: Sensing Hope in Post-Earthquake Japan." Unpublished paper presented at the American Anthropological Association. Retrieved from https://supervalentthought.com/sensing-precarity-allison-stewart-garcia-berlant-mclean-biehl (accessed June 2, 2015).

Arnold, Dennis, and Joseph Bongiovi. 2013. "Precarious, Informalizing, and Flexible Work: Transforming Concepts and Understandings." *American Behavioral Scientist*, 57(3): 289–308.

Beck, Ulrich. 1999. *World Risk Society*. Cambridge: Polity Press.

Bergeron, Suzanne, and Jyoti Puri. 2012. "Sexuality Between State and Class: An Introduction." *Rethinking Marxism*, 24(4): 491–498.

Bernhardt, Annette. 2014. *Labor Standards and the Reorganization of Work: Gaps in Data and Research*. University of California, Institute for Research on Labor and Employment, Working Paper #100-14.

Boyer, Robert. 1998. "Wage Determination and Distribution in Japan by Toshiaki Tachibanaki." *Journal of Japanese Studies*, 24(1): 155–160.

Brinton, Mary. 2011. *Lost in Transition: Youth, Work, and Instability in Postindustrial Japan*. Cambridge: Cambridge University Press.

Butler, Judith. 2006. *Precarious Life: The Power of Mourning and Violence*. London: Verso.

Chang, Dae-Oup. 2011. "The Rise of East Asia and Classes of Informal Labour." Unpublished paper presented at the Sawyer Seminar, University of North Carolina, February 14.

Cook, Emma. 2013. "Expectations of Failure: Maturity and Masculinity for Freeters in Contemporary Japan." *Social Science Japan Journal*, 16(1): 29–43.

Dasgupta, Romit. 2003. "Creating Corporate Warriors: The 'Salaryman' and Masculinity in Japan." In *Asian Masculinities: The Meaning and Practice of Manhood in China and Japan*. K. Louie and M. Low, eds. London: Curzon Press, pp. 118–134.

Dasgupta, Romit. 2012. *Re-reading the Salaryman in Japan: Crafting Masculinities*. London: Routledge.

Endo, Chikako. 2006. "Review of Richard F. Calichman, *Contemporary Japanese Thought*. New York: Columbia University Press, 2005." H-US-Japan, H-Net Reviews. Retrieved from www.h-net.org/reviews/showrev.php?id=11518 (accessed January 27, 2014).

Fourcade, Marion. 2010. "The Problem of Embodiment in the Sociology of Knowledge: Afterword to the Special Issue on Knowledge in Practice." *Qualitative Sociology*, 33(4): 569–574.

Fraser, Nancy, and Linda Gordon. 1994. "A Genealogy of Dependency: Tracing a Keyword of the US Welfare State." *Signs*, 19(2): 309–336.

Gill, Tom. 2003. "When Pillars Evaporate: Structuring Masculinity in the Japanese Margins." In *Men and Masculinities in Contemporary Japan: Dislocating the Salaryman Doxa*. J. Roberson and N. Suzuki, eds. London: RoutledgeCurzon, pp. 144–161.

Gordon, Andrew. 1985. *The Evolution of Labor Relations in Japan*. Cambridge, MA: Harvard University Press.

Gottfried, Heidi. 2009. "Japan: The Reproductive Bargain and the Making of Precarious Employment." Chapter 5 in *Gender and the Contours of Precarious Employment*. L. Vosko, I. Campbell, and M. MacDonald, eds. London: Routledge.

Gottfried, Heidi. 2013. *Gender, Work and Economy: Unpacking the Global Economy*. Cambridge: Polity Press.

Gottfried, Heidi. 2014a. "Rescaling Gender and Labor Politics: Geographies of Power and Resistance." In *Transnational Spaces and Gender* [*Transnationale Räume und Geschlecht*]. B. Riegraf and J. Gruhlich, eds. Munster: Verlag, pp. 176–192.

Gottfried, Heidi. 2014b. "Precarious Work in Japan: Old Forms, New Risks?" *Journal of Contemporary Asia*, 44(3): 464–478.

Gottfried, Heidi. 2015. *The Reproductive Bargain: Deciphering the Enigma of Japanese Capitalism*. Leiden: Brill.

Grossberg, Lawrence. 2013. "Culture." *Rethinking Marxism*, 25(4): 456–462.

Haney, Lynne, and Miranda March. 2005. "Married Fathers and Caring Daddies: Welfare Reform and the Discursive Politics of Paternity." *Social Problems*, 50(4): 461–481.

Hazama, Hiroshi. 1997. *The History of Labour Management in Japan*. London: Macmillan.

Hirakawa, Hiroko. 1995. "Inverted Orientalism and the Discursive Construction of Sexual Harassment: A Study of Mass Media and Feminist Representation of Sexual Harassment in Japan." PhD dissertation, Purdue University, West Lafayette, IN.

Imai, Jun. 2011. *The Transformation of Japanese Employment Relations: Reform Without Labor*. Basingstoke: Palgrave Macmillan.

Japan Institute for Labour Policy and Training (JILPT). 2012. *Labor Situation in Japan and Its Analysis*. Tokyo: JILPT.

Japan Institute for Labour Policy and Training (JILPT). 2013. *Japanese Working Life Profile 2012/2013*. Tokyo: JILPT.

Japan Institute for Labour Policy and Training (JILPT). 2014. *Japanese Working Life Profile 2013/2014*. Tokyo: JILPT.

Japan Institute for Labour Policy and Training (JILPT). 2015. *Labor Situation in Japan and its Analysis*. Tokyo: JILPT.

Kalleberg, Arne. 2011. *Good Jobs, Bad Jobs*. New York: Russell Sage Foundation.

Kalleberg, Arne, and Kevin Hewison. 2013. "Precarious Work and the Challenge for Asia." *American Behavioral Scientist*, 57(3): 271–288.

Krishnan, Sneha. 2014. "Book Review: *Precarious Japan* by Anne Allison." *LSE Review of Books*. Retrieved from http://blogs.lse.ac.uk/lsereviewofbooks/2014/01/29/book-review-precarious-japan-by-anne-allison (accessed February 23, 2014).

Kumazawa, Makoto. 1996. *Portraits of the Japanese Workplace: Labor Movements, Workers, and Managers*. Boulder, CO: Westview Press.

Kurokawa, Michiyo. 1995. "Japan." *Bulletin of Comparative Labour Relations*, 30: 45–90.

Lorey, Isabell. 2011. "Governmental Precarization." Retrieved from http://eipcp.net/transversal/0811/lorey/en (accessed June 15, 2015).

Ministry of Health, Labour and Welfare. 2006. *Basic Survey on Wage Structure*. Tokyo: Ministry of Health, Labour and Welfare.

O'Day, Robin. 2012. "Review of: M. Brinton, *Lost in Transition: Youth, Work, and Instability in Postindustrial Japan*." *Book Reviews*, 85(1). Available at: www.pacificaffairs.ubc.ca

Roberson, James, and Nobue Suzuki. 2003. "Introduction." In *Men and Masculinities in Contemporary Japan: Dislocating the Salaryman Doxa*. J. Roberson and N. Suzuki, eds. London: RoutledgeCurzon, pp. 1–19.

Rosenbaum, Roman. 2014. "Towards an Introduction: Japan's Literature of Precarity." Chapter 1 in *Visions of Precarity in Japanese Popular Culture*. K.I. Weickgenannt and R. Rosenbaum, eds. Oxon: Routledge.

Shire, Karen. 2012. "The Work-Welfare Nexus in Post-Disaster Japan: Deepening Social Risks or New Opportunities for a Better Work-Life Balance?" Paper presented at the DFG Research Training Group 1613.

Slater, David H. 2009. "The Making of Japan's New Working Class: 'Freeters' and the Progression from Middle School to the Labor Market." In *Social Class in Contemporary Japan: Structures, Socialization and Strategies*. H. Ishida and D.H. Slater, eds. New York: Routledge, pp. 103–115.

Slater, David H., and Patrick W. Galbraith. 2011. "Re-Narrating Social Class and Masculinity in Neoliberal Japan: An Examination of the Media Coverage of the 'Akihabara

Incident' of 2008." *Journal of Contemporary Japanese Studies*, 7. Retrieved from www. japanesestudies.org.uk/articles/2011/SlaterGalbraith.html (accessed August 12, 2012).

Standing, Guy. 2011. *The Precariat: The New Dangerous Class*. London: Bloomsbury Academic.

Tabata, Hirokuni. 1998. "Community and Efficiency in the Japanese Firm." *Social Science Japan Journal*, 1(2): 199–215.

Uno, Kathleen. 1993. "The Death of 'Good Wife, Wise Mother?'" In *Postwar Japan as History*. A. Gordon, ed. Berkeley, CA: University of California Press, pp. 293–322.

Yoda, Tomiko. 2006. "The Rise and Fall of Maternal Society: Gender, Labor, and Capital in Contemporary Japan." In *Japan After Japan: Social and Cultural Life from the Recessionary 1990s to the Present*. T. Yoda and H. Harootunian, eds. Durham, NC: Duke University Press, pp. 239–274.

4 Youth "volunteers," unemployment, and international action in Pakistan's health sector

"I need money, that's the only reason I do it"

Svea Closser

We're not that well off [*ham inti set nahin hain*]. My father has a salary of 4,500 rupees a month [about US$65], and I have five brothers and sisters. My mother does sewing at home too, for money. When I was in school, some of my school expenses would be covered by my polio vaccination work.

Polio payments are becoming very small [because of inflation], things are getting expensive, and 150 rupees [about US$2, the daily stipend for polio work] isn't much. Before, things were affordable, but now everything is becoming expensive. Also, I have to pay transport to get to work now. I don't have the strength to walk that far [*itni dur jaane ki himat nahin hoti*]—I have to come home and do my housework too. So I pay transport to get there, I pay transport to get back, you have to eat lunch [. . .] in the end I only make 80 rupees [US$1.25] a day.

I need money, that's the only reason I do it.

(Female teen "volunteer" on polio eradication campaigns, Karachi, Pakistan)

From a humanitarian perspective, [polio] eradication provides the ultimate in health equity and social justice, bringing identical and universal benefits to every person globally.

(Global Polio Eradication Initiative leadership [Aylward et al. 2000: 286])

The Global Polio Eradication Initiative aims to rid the world completely and permanently of polio—an untreatable viral disease that can cause paralysis and death—through the vaccination of every child on earth. This goal, often framed in language of social justice, is both difficult and ambitious. Delivering a polio vaccine to every child on the globe, multiple times, is a huge task— one probably unprecedented in public health history. It has drawn on the labor of tens of millions of people across the world, people who visit nearly every community in sub-Saharan Africa (SSA) and South Asia—often going door-to-door. These workers vaccinate children, keep records of who was missed, and return again and again in an attempt to stop the spread of the poliovirus. These workers—referred to as "volunteers"—generally make only a few dollars a day, commonly less than the minimum wage in the country where they are working (Closser et al. 2014). This practice—common in the global health industry—allows the Global Polio Eradication Initiative to frame its labor practices to donors as "community participation," while in fact taking advantage of conditions of high unemployment to extract the labor of the vulnerable poor, particularly women, at low cost. Thus, this global health project, one motivated in large part by social justice, entrenches inequality through its labor practices even as it alleviates inequality in the narrow realm of polio transmission.

I explore in this chapter the political economic context—of both global health institutions and the country of Pakistan—that results in underemployed women in Pakistan taking up "volunteer" work for polio eradication. In June 2011, I conducted in-depth semi-structured interviews with seven "volunteers," young women in Karachi and Hyderabad, Sindh, who ranged in age from 11 to 30. (Officially, volunteers are supposed to be at least high-school age, but despite the best efforts of national planners—who feel that children are too immature for the work—the low wage sometimes results in only children being willing to take the job.) For background information, I also drew on interviews and ethnographic fieldwork on polio eradication carried out in more than 12 non-contiguous months between 2007 and 2011 in Islamabad, Geneva, and a city in the Pakistani Punjab. The specifics of polio work in these cities has changed since the situation described in this chapter in 2011. This chapter should not be viewed as a depiction of current events but as a case study of how local labor markets and global health projects interact.

The context of female youth labor in Pakistan

Polio eradication officials strongly prefer to fill their "volunteer" positions with women because, unlike men, women are welcome to enter strangers' homes in Pakistan, and can vaccinate newborns during the first 40 days of their lives, when they are often kept secluded inside the house. "Volunteer" work for women in Karachi and Hyderabad fits into the context of urban women's work more generally. Figures on women's labor participation and unemployment rates in Pakistan historically have been unreliable and, although they have improved in the last 20 years, continue to be highly contested (Kazi and Raza 1989; Khan 2007). Trustworthy statistics on unemployment rates for women in Pakistan are difficult to obtain for several reasons.

First, regardless of financial need, women who do not work do not generally consider themselves "unemployed." Since the dominant cultural ideal is that women do not work for pay, but rather are supported by their husbands or fathers, women are unlikely to describe themselves as unemployed even if they are in need of money and in search of work. Official employment estimates that women work much less than men are also difficult to assess. As Anita Weiss (1992: 74) explains, in Pakistan "concerns over traditional notions of propriety have not prevented women from working for pay; instead, they have often simply prevented women and their families from admitting that they engage in such work."

Second, much of women's labor participation takes place inside the home, and is under-reported and relatively invisible. Contributing to this invisibility at the policy level is the fact that many surveys of labor-force participation in Pakistan are carried out by men, who are less likely than women to obtain accurate answers from female respondents (Kazi and Raza 1989; Weiss 1992; Khan 2007). The gender segregation and ideals of seclusion prevalent in Pakistani society mean that women prefer not to talk to unrelated men, and are often reluctant to invite them into their home to conduct an interview—making it difficult for male enumerators to obtain accurate survey responses from women.

A survey of female young people carried out in the early 2000s showed that, by age 25, well over 40% of respondents had participated in paid work (Lloyd and Grant 2004).[1] The government's official estimates of labor participation for women are 25% for Pakistan as a whole and just 8.4% for urban women in the province of Sindh, where the interviews presented in this chapter were conducted. But these figures must be taken with a pinch of salt: because of the methodological problems described above, they are almost surely underestimates.

Unlike in many other global cities (Sassen 2000), women in Karachi and Hyderabad do not form much of the service sector, even in the informal sector, outside the home. Ideals of seclusion mean that jobs like secretarial work, garment making, and cleaning—jobs held by women in many global cities—are primarily performed by men (Makino 2012). Employment options for women in urban Sindh are thus severely limited.

Official unemployment rates are striking: nearly 9% for women across Pakistan, and a full 21% for urban women in Sindh (Government of Sindh 2013). But because many unemployed women do not think of themselves as unemployed, so-called inactivity rates (accounting for women who are officially not looking for work) are also important to consider. In 2008, the best estimates were that 70% of women in Pakistan were "inactive" (Government of Pakistan Ministry of Labour and Manpower 2009). But these inactivity rates likely hide much home-based work and much unemployment.

Ethnographic research can provide better, albeit context-specific, information. From an anthropological perspective, the most comprehensive and reliable— although dated—research on urban women's work in Pakistan is a book by Anita Weiss on women's work in Lahore, based on fieldwork carried out in the late 1980s. Lahore is in the Punjab, a province north of Sindh. In Weiss's (1992) random sample of 100 women in Lahore's Walled City, 33 worked for pay, and another 34 said they wanted to work, but did not have opportunities to do so. This unemployment rate in the 30% range is much higher (and likely more accurate) than official estimates.

My own observations in the Punjab 20 years later support Weiss's high estimates of female labor participation. Beginning in 2005 I spent nearly two years living in a lower-middle-class area of Rawalpindi. Most of the women I knew who were my age (I was in my late 20s at the time) worked for pay, although few would have described themselves as employed. Nearly all of them worked from home, and virtually none had full-time employment. All chose work that they could do from home: some tutored neighborhood children after school; some sewed clothes for pay; some made quilts for sale; some prepared food or other items to be sold at male family members' shops. Potential income-generating opportunities were a frequent topic of conversation.

Women's labor-force participation in Pakistan is heavily shaped by ideals of gender segregation—women much prefer work that they can do from home, remaining in the female-dominated domestic sphere rather than the male-dominated public sphere. Women in Pakistan frequently frame the desirability of seclusion and gender segregation using the term *purdah*—literally, "curtain"—with reference to the curtain that Muhammad's wives used to shield their private lives from outsiders. But the practice of gender segregation in South Asia is at least as powerfully shaped by class and ethnicity as by religion (Jalal 1991). Hindu women in North India also structure their lives according to the concept of *purdah*, although the boundaries are slightly different (Gold 1994). High-class professions like doctor (and prime minister) are relatively open to women in Pakistan. At the same time, lower-class professions are largely closed to women. Jobs considered "women's work" in the West, like nursing and secretarial work, are occupied by men (c.f. Papanek 1973).

Most urban lower- and middle-class women who are employed work in their own homes. While such work allows women to maintain ideals of seclusion, rates of pay are low and work can be inconsistent. Often they perform the same work as men for lower pay, and must rely on middlemen who drop off raw materials and pick up the finished work (Kazi and Raza 1989; Weiss 1992). Such work, which occurs outside the so-called formal economy, is unregulated. The implications of this are far-reaching. For example, the International Labour Organization (ILO) advocates for what it calls "decent work," where workers earn at least minimum wages for the country they live in, have the right to organize and have their demands heard, and receive recognition, respect, and legal protections (ILO 2015). Unregulated work, such as working at home for an unregistered business or "volunteering" on a polio campaign, generally comprises non-decent jobs.

Of the women known to be working in Pakistan, the vast majority are outside the formal sector. In 2008, it was estimated that only one-fifth of working women in Pakistan were working in waged or salaried employment.[2] Since the previous survey was done in 2000 more women had joined the labor force, but there had been a decrease in decent employment (Government of Pakistan Ministry of Labour and Manpower 2009: 19).

Although the wages that women earn outside the formal sector are low, the contributions to family finances may be nonetheless significant. One survey of home-based female workers in Karachi carried out in 1987 found that, on average, these workers' salaries made up one-third of family incomes and, in the lowest income bracket, they contributed more than half of the family income (Kazi and Raza 1989).

The dynamics at play here are long-standing across South Asia, and were beautifully analyzed by Maria Mies in her writing about female home-based work in South India in the late 1970s. As Mies (1981: 487) explains, the myth that women do not work for pay allows them to be exploited: "only as long as the mystification is maintained that they are non-workers, or dependents, is it possible to keep them unorganized, atomized, and to pay them wages which are much lower than even the wages of agricultural laborers." Female workers in the informal economy in Pakistan today, like female workers in the informal economy in South India 40 years ago, are well aware that their labor is undervalued and exploited. But the ideal of females as non-workers makes it difficult to enact change.

The larger political-economic context

The political economy of modern Pakistan also shapes the conditions of informal employment. In the early 2000s under Prime Minister Shaukat Aziz, Pakistan embraced neoliberal policies, and experienced robust economic growth. In 2007, however, in the face of high inflation, high unemployment, and shortages of basic commodities such as water and electricity, protests and unrest led to the downfall of Musharraf and Aziz's government. After the elections, the new leaders inherited an insolvent government and inflation rates reaching 20%. The International Monetary Fund (IMF) loan of US$11.3 billion, which they accepted to keep the government afloat, carried conditions that further entrenched the neoliberal agenda (Akhtar 2013). At the same time, post-9/11, the United States has given on average US$2 billion in aid a year to Pakistan, the vast majority of which goes to the military. This has contributed to the further entrenchment of military power in a state that is not particularly responsive to the needs of its own citizens (Zaidi 2011; Akhtar 2013).

Because services for the population as a whole are not a priority for the elites that run the country, modern Pakistan is a lower-middle-income country with health and education indicators that lag behind other countries of similar incomes (World Bank 2014). Eighty-six out of every thousand children die before they reach the age of five—a child mortality rate that is 50% higher than that of India (WHO 2014a).

Karachi and Hyderabad, like many other cities in Pakistan, are characterized by a highly unequal distribution of wealth, uncertainty due to armed conflict (including the War on Terror), and a lack of investment in urban infrastructure and services. Water shortages, fuel shortages, high inflation, and rapidly rising food and fuel prices provide the backdrop to women's need for additional income. Inflation rates when these interviews were carried out were nearly 14%, and nearly 18% for food items (State Bank of Pakistan Statistics and DWH Department 2011).

The global polio eradication initiative

The Global Polio Eradication Initiative was started in 1988, with the goal of eradicating polio entirely from the earth by 2000. This very ambitious and difficult goal was not achieved and, in 2016, polio persists stubbornly in Afghanistan and Pakistan, with ongoing outbreaks in other parts of the globe. In the 15 years since the polio eradication initiative's first missed goal, it has scaled up considerably, chasing the elusive goal of the extinction of the poliovirus. Historically, polio eradication had been a partnership between the World Health Organization (WHO), United Nations Children's Fund (UNICEF), the United States Centers for Disease Control and Prevention (CDC), and Rotary International, as a way of taking on a project too large, complex, and expensive for any one of these entities to handle on its own. In 2007, the Bill and Melinda Gates Foundation, a relatively new player on the global health scene, embraced polio eradication as a central goal, and its deep pockets pushed the polio eradication effort to a new level of intensity. More than US$10 billion has already been spent on the effort to eradicate polio, and now polio eradication activities cost over US$1 billion a year (WHO 2014b).

There are three countries—Nigeria, Afghanistan, and Pakistan—where polio transmission has never been stopped. Of the 414 cases of polio in the world in 2014, 328 of them were in Pakistan.[3] Pakistan is part of what polio eradication's oversight board calls the "hard core" of polio transmission, and it is where much of the money spent is concentrated (Independent Monitoring Board of the Global Polio Eradication Initiative 2013). It has been estimated that expenditures for polio eradication in Pakistan alone will be nearly US$700 million over the next five years (WHO 2014b). The lion's share of this money in Pakistan is spent on delivering the polio vaccine door-to-door. Oral polio vaccine, which can be administered by laypeople, is delivered to every child in Pakistan aged under five multiple times a year. In areas like Karachi, where polio transmission is ongoing, there are as many as 12 campaigns a year, in the hope that increasing the intensity of the vaccination effort will lead to a critical mass of immune children and an end to polio transmission. With huge requirements in terms of vaccine, logistics, and supervision, these campaigns are expensive. For example, in 2014, the requested budget for vaccine costs for Pakistan was US$58 million; the requested budget for carrying out campaigns was approximately US$38 million; disease surveillance was approximately US$5 million; "technical assistance" (mostly international consultants) was approximately US$20 million; and "social mobilization" (getting people to accept the vaccine) was approximately US$26 million—making a total budget of US$146 million (Global Polio Eradication Initiative 2015).[4] In 2014 the people who did the actual work of vaccinating were paid around US$2.50 a day, an amount well below the monthly legal minimum wage of 12,000 rupees (about US$125). (The international consultants who worked on the project, in contrast, were paid well over US$10,000 a month.)

Polio is a disease with fecal-oral transmission that does not exist in wealthy countries like the United States because of effective sanitation and reasonably high levels of routine immunization coverage. There are two potential ways to eliminate the disease: achieve good sanitation and immunization everywhere, or focus on a resource-intensive but relatively short-term program of repeated vaccination campaigns to eliminate the disease without making larger improvements to infrastructure. Polio eradication's leaders made the decision to pour funding, oversight, and international pressure into campaigns rather than into strengthening health systems, improving sanitation infrastructure, or improving vaccination coverage for all childhood immunizations (Closser et al. 2014). Rather than spending money on building strong health systems with professional workers, they rely on campaigns staffed largely by unskilled "volunteers" paid less than Pakistan's official minimum wage (as explained above). The rationale for this decision is that building sanitation and health systems is too slow, too difficult, and too expensive; in contrast, vaccination campaigns seem easy (Closser 2010). Thus, the door-to-door work of vaccinating children is firmly embedded in the political economy of global health projects. The Global Polio Eradication Initiative conducts as many as 12 door-to-door polio vaccination campaigns per year in Karachi—a practice that has become a necessity because of the reluctance of both the Pakistani government and international aid agencies to invest meaningfully in well-supported systems of primary healthcare.

India, which saw its last polio case in 2010, offers an instructive counter-example. Faced with stubborn polio transmission in the states of Bihar and Uttar Pradesh—areas where health systems were weak—Indian and international planners decided that, beyond just focusing on campaigns, they would build broad communication messaging for health around the polio program and, insofar as was possible, use it to improve other health services (Closser et al. 2014). This strategy helped to allay fears about polio campaigns, and contributed to the end of polio in India. But in Pakistan, where health systems are weak, a heavy schedule of polio campaigns has continued for 15 years—politicizing the project and drawing criticism. While refusal rates for the polio vaccine in Pakistan are very low—the best estimates are less than 1% overall—some parents who accept the polio vaccine in general have started refusing it during campaigns, frustrated that, while the polio vaccine is delivered to their doorstep multiple times a year, more pressing health issues remain unaddressed (Closser et al. 2015). As I will discuss later, the high profile of campaigns has made them a target for militants. Many, even those within polio eradication, feel that this reliance on campaigns has been a mistake. At a conference in 2015, a high-ranking polio eradication official said he regretted the "excesses" of having too many campaigns, and that building a polio eradication project that was more integrated with the health system would have been better. But still, now as for the past 20 years, the work of eliminating polio falls to Pakistan's "volunteers."

Volunteerism in global health

Polio eradication is not the only global health entity to rely on underpaid "volunteers" to do the work of healthcare provision. Many of the world's largest public health initiatives rely on underemployed local people to provide unpaid or very low-paid labor (Maes 2010). In an era of unprecedented funding for global health (Ravishankar et al. 2009), relying on poorly-paid or unpaid labor to deliver critical health interventions in poor countries is a standard practice.

Polio eradication policymakers rely on "volunteer" labor because they can, and because it saves money. A senior polio eradication official I interviewed in Geneva in 2007 explained that the small stipend given to workers "seems to be enough" to get them to work. Another polio eradication official for the Middle East region, based in Cairo, dismissed Pakistani officials' assertion that "volunteers" should be paid more at a 2007 meeting, saying: "We recently increased pay by 50% [to less than US$3 per day], but did we get an increase in quality to compensate for this? I have my doubts."

Aside from simple financial convenience, polio eradication leadership has long used the supposed commitment of its "volunteers" as a strategy to convince potential donors that there is overwhelming community support for polio eradication's activities. In the political economy of health aid, the enthusiasm of wealthy donors for volunteer activities is in many ways more important than the enthusiasm (or lack thereof) of the "volunteers" themselves. Senior officials at organizations affiliated with polio eradication are well versed in the party line. A senior official at the United States Agency for International Development (USAID) said in an interview for a movie about polio eradication: "Never before have there been millions of volunteers, just volunteering their time to vaccinate, to promote, to advocate, to mobilize, people in the remotest areas of the world" (Thigpen 2004). Similarly, Bruce Aylward, who runs polio eradication at the WHO, said in an interview on Canadian radio:

> The trick to overcoming the logistical challenges—and this is I think the greatest legacy of the polio program—was that we actually put the intervention, the oral polio vaccine, in the hands of the parents and volunteers from the communities themselves. We estimate at the peak of our operation, probably 20 million people, some of the poorest most destitute people in the world, were a key part of this program delivering this vaccine to their children [...] And what was really so encouraging, Barry, and I think is the bigger lesson from polio, is that when you put the resources and the tools in the hands of these people they will do extraordinary things to protect their children.
>
> (cited in Dworkin 2007)

In the early years of polio eradication, when campaigns took place in the Americas, true volunteers—namely wealthy elites, many of them members of Rotary—provided vaccinations. In the Americas, the number of campaigns required was low, and volunteering for polio eradication was a matter of a few days a year—a situation starkly different from that in Pakistan today, where campaigns can employ a given volunteer upwards of 70 days a year (Daniel and Robbins 1999).

The money that "volunteers" receive in Pakistan—which, at the time of my interviews, was less than US$3 per day—is officially not pay, but a stipend to defray the costs of lunch or transportation. This discourse of volunteerism continues because it serves several goals: it keeps costs down, and it allows donors to believe both that there is strong community commitment to polio vaccination and that their dollars represent a philanthropy deal. For example, ten years ago polio eradication officials wrote in their Strategic Plan that "polio endemic countries will have contributed volunteer time worth at least $2.35 billion for polio eradication activities between 1988 and 2005" (WHO 2003).

The "volunteer" label exists in a political economy of health aid where expatriate labor and local labor are differently valued, both reflecting and reproducing the inequalities between powerful aid organizations and those they intend to help (Fassin 2007). The fact that global health projects exist primarily to "do good"—eradicating polio, after all, would be a wonderful thing—means that their moral discourse often silences critique (Redfield 2006; Tichtin 2006). Reliance on underpaid or unpaid labor is frequently justified with the argument that diseases of poverty are crises demanding exceptional action from citizens and governments alike (Maes 2012).

The use of volunteerism in global health is shaped by a perceived context of resource scarcity. Not paying wages to people who provide frontline health labor is often rationalized as economically imperative when it comes to designing and implementing "sustainable" health projects in resource-poor settings (Akintola 2008; Swidler and Watkins 2009). But the perception of scarce resources for frontline health labor exists in the context of historically unprecedented funding for global health (Ravishankar et al. 2009). Resources for paying frontline labor are not simply scarce and unsustainable: even as frontline salaries remain low, donors have increased funding to medical technologies, supplies, infrastructure, and higher-level personnel (Drager et al. 2006; Biehl 2007; Ooms et al. 2007; Swidler and Watkins 2009). Instead, in resource-poor countries, wealthy global health and development agencies frequently depend on "volunteers." This term is complicated across the world: volunteers, as workers without pay, are part of an unstable workforce. Felicitas Becker (2015: 116) explains that, globally, the "slippage in status connotations of unpaid work between 'posh' and precarious highlights how context-dependent and unstable the status of the unpaid worker can be." In resource-poor contexts, the word "volunteer" hides more than it reveals: volunteers for global health and development projects are frequently impoverished, using "volunteer" work as part of a strategy to scrape together a living. Frequently, high levels of unemployment push people towards "volunteer-ing" for small food stipends or in the hope of paid employment later (Maes 2012; Kalofonos 2014; Becker 2015; Brown and Prince 2015). The term "volunteer," Becker (2015: 128) explains, "look[s] clear on paper, whatever the indeterminacies on the ground"—making it useful when selling programs to donors. International development discourse has constructed female young people in particular as "virtuous victims"—morally pure and lacking agency, yet somehow able to mobilize for health and development goals (Chant 2016). So volunteerism—female volunteerism in particular—is also justified with narratives that construct volunteers as benefiting from their service (Glenton et al. 2010). Global narratives of female participation in health programs construct female workers as improving health and empowering themselves simultaneously as a result of their participation in health work (Ramirez-Valles 1998). The female workers themselves, though, seldom see their employment in quite these terms (Maes et al. 2010; Closser and Jooma 2013). Detailed ethnographic research on other health-worker programs in Pakistan show that women's relationship with their work is complicated—it empowers them in certain ways (giving them some greater freedom to move, for example) and limits them in others (Mumtaz et al. 2013). Paid female community health workers in Pakistan find that their position at the bottom of the nearly all-male health system hierarchy is one in which they have very little power (Mumtaz et al. 2003; Khan 2011).

The language of volunteering in Pakistan occurs in a context, then, not only of community engagement but also of hierarchy—the rigid hierarchy of the Pakistani health system, and above that the moneyed world of international consultants and donors. It is in this context that "volunteerism" is valued for its cost-efficiency. The fact that such volunteers are so frequently women led Katherine Brickell and Sylvia Chant (2010: 146) to describe the female "altruistic burden" as "one of the deepest bastions of gender inequality." Chant (2008: 186) argues that increasing female volunteerism may be "a new and deeper form of female exploitation."

Polio eradication's volunteers

Despite the rhetoric, nearly everyone involved in polio eradication in Pakistan, from high-level officials down to the volunteers themselves, understands that the money offered to volunteers functions as pay. Polio work is not easy: it requires substantial walking, sometimes in temperatures above 45°C (113°F), and meticulous recordkeeping regarding what children live where, and whether they have been immunized. "Volunteers" that are found to have missed children are severely reprimanded by supervisors, as will be discussed later. Within Pakistan, at the district level, there is strong support for increased pay for volunteers. I interviewed 33 people who supervised volunteer work at the district level in districts across Pakistan in 2011, and they were unanimous in their assertion that the biggest problem in motivating ground-level workers was low pay. Supervisors and Area in Charges repeatedly mentioned in interviews that they found it extremely difficult to attract and retain qualified workers given the low rate of pay and the late delivery of pay. One supervisor told me: "We have to find volunteers however we can, asking our family members, our neighbors. It feels like we're beggars." Another said: "How do you expect to have workers that look presentable, with nice clothes, when they're being paid 150 rupees a day?"

The seven women I focus on in this chapter are young female volunteers in Karachi and Hyderabad, megacities in the Pakistani province of Sindh. All volunteered for one, or a combination, of three reasons: (1) a relative with a low-level job in the government health system needed reliable workers for the polio campaign; (2) the volunteers hoped that working for the polio campaign would lead to a job in the health sector; and/or (3) the volunteers' families faced severe poverty.

Volunteering to help a relative

Four of the women I interviewed volunteered because they had a relative who was a Lady Health Worker—a female community health worker. These Lady Health Workers were responsible for finding volunteers to work on the polio campaign, and they were having trouble finding people that were reliable and hardworking—or, for the extremely low pay offered, they were having trouble finding any volunteers at all. These Lady Health Workers were stressed because they were held responsible for the vaccination coverage of the volunteers they supervised. So some women came to the aid of their Lady Health Worker relative by volunteering. One volunteer in her late teens explained:

> My mother, she works as an area in charge. And the teams she used to have, they didn't take the work seriously [*la-parvaai se kaam karte thhe*], and it was incredibly worrying for her, and it was really hard, so that's why I started, so, you know, I'll work really well? She won't have any worries. So that's why I started.

When I asked if she was personally interested in the work, this volunteer laughed. "No, I wasn't interested [*nahin, shoq nahin thha*]."

Volunteering in the hope of a job in the health sector

Many women volunteered because they hoped to get a job in the health sector, usually the low-level Lady Health Worker position. Usually, they already had the credentials for such work, and hoped that working on the polio campaign would show supervisors the quality of their work, and lead to a permanent job. While this tactic was sometimes successful—many Lady Health Workers did polio work before getting this more permanent position—not all volunteers who hoped for jobs obtained them. Compared to the numbers of volunteers, numbers of Lady Heath Worker jobs were low, and most of them were already filled.

Volunteering because of poverty

Many women took up polio "volunteering" as a way of earning a bit of extra money in the context of desperate straits at home. Many women spoke of entering the work using the Urdu word *majboori*, which carries connotations of having no choice, of being obligated or forced into something. One woman in her late 20s had recently stopped volunteering. She explained:

> I started out of *majboori* [obligation], because these days there are so many problems, and you can't survive on one income. My husband doesn't have a

government job; he's a day wage laborer. In a month, he earns 5,000 or 5,500 rupees [about US$75]. It's hard to live on that. So I said, if I earn a little money, it will be good. But you have to walk so far, and you earn so little money, and that so long after you've done the work. That's why I refused—I said I'm not doing it any more.

The woman I quoted in the epigraph for this article, in her late teens, also volunteered because of poverty. Like many other volunteers, this woman also opportunistically engaged in other moneymaking activities, like sewing piece-work, that could be done from the home. Her mother worked on sewing nearly full-time, and she helped out. But despite the fact that many volunteers took up the work because they desperately needed money, women repeatedly mentioned that the take-home pay for polio work was extremely low—even in the context of other low-paid work.

The problem of low pay

Workers voiced their frustrations over low pay in no uncertain terms. Sentiments including "the pay is nothing" and "polio pay is much too low" were voiced in nearly every interview. Respondents pointed out that, because of inflation (which was around 15% at the time of this research), their low pay was constantly becoming even lower in real terms.[5] Again and again, interviewees compared polio pay to the going rates of pay for unskilled labor (*mazdoori*). Estimates of the going rates of pay for unskilled labor were over 500 rupees (US$8) per day in Karachi. Respondents repeatedly pointed out that, if unskilled laborers earned that much, polio workers—who were required to have some education—should reasonably be expected to make a bit more than that. For the vast majority of the workers interviewed, frustration over low pay was not an abstract problem or a problem of status. I asked volunteers about the size of their families and the availability of income, and the majority of them were living at a rate of income around, or in many cases well below, US$2 a day per person. Volunteers repeatedly mentioned extreme difficulty in covering unforeseen expenses such as medical treatment for a child's illness or an adult's injury. One volunteer in her late 20s explained:

> I need the money, right? If I made better money, I could get things for my children. I think it's better than complaining about your fate to work hard. Feed your children; raise them well. But when after working so hard you get so little money, your heart breaks. And you don't want to do that work. It's the truth.

In the normal course of doing polio work, volunteers incurred a number of expenses that were not reimbursed, making their pay even lower in real terms.

Nearly as upsetting to workers as the low rate of pay (and mentioned almost as frequently in interviews) was the problem of late or missing pay. First, pay for polio rounds often came late—in the best-case scenario, a month late and, in worse cases, after as many as four months. A woman in her late teens explained:

> What's the benefit to all that work? When we work so hard. Fine, if it's one week late, no problem. It used to be like that, one week late, and then we would get our money. But not anymore [...] It doesn't come, it doesn't come, it doesn't come, and a person even forgets, you know? It seems like there's no chance it will ever arrive. Our hope dies [...] We look and look for it, we wait and wait for it, and then when we've completely given up hope and forgotten about it, it arrives.

Relationships with community members and supervisors

While most people accepted polio vaccination for their children nearly automatically, a minority did not. The response of the public formed an important part of respondents' work experience. A volunteer in her late teens explained:

> It feels so bad when people don't give a good response. People don't understand that we come because we have to. They say: "We haven't given our children polio drops, they haven't gotten polio, why are you doing this?" And then they say there's some family planning thing in the drops, like it will keep their children from having kids, and they say: Where are you from? Who sent you? But when a good, polite person comes to the door, it's so nice! It makes your heart so happy.

In Karachi, a town with severe law-and-order problems (several times I had to cancel interviews because gunfights in the streets prevented me from reaching my destination), work was particularly difficult. One woman in Karachi described working all day and being unable to get so much as a glass of water from the houses she visited. Another said: "We don't trust them, they don't trust us."

Not only community members, but also workers' own supervisors could create negative work experiences. Not all experiences with supervisors were equal, but in all cases negative feedback from supervisors was abundant. A significant number of workers were hurt and angry about their experiences. One volunteer said that: "150 rupees per day would just about cover my trouble for the abuse I get. It doesn't begin to cover working in the heat."

Competing responsibilities at home

Many women complained about the amount of time they were required to spend away from their families. Married women mentioned logistical challenges in being away from the house all day. On polio days, nearly all married women got up early in the morning, around 4 a.m., to cook, do housework, and get their children ready before they left for polio work. For women with small children, being away all day could be a challenge. Several of the women who quit doing polio work cited family responsibilities as an issue. A former volunteer in her late 20s explained:

> If I get paid good money, I'll do it. But when we're paid so little, how can I do it? It takes almost the whole day. I leave in the morning; I'm gone the whole day. I have small children. Leave your children behind, do all the housework quickly early in the morning. Leave the house by 9 or 10. So it creates problems for me, and I'm not doing it.
>
> You know, if the money was good, I could deal with the other issues. But there's so little money, and at home I have so many responsibilities, that's why I refused, I said I'm not doing it any more. But if I were to make good money, why wouldn't I do it? Obviously, I need the money! And my work is good—I work carefully, actually, nobody would have to check my work.

This woman also did handiwork at home. If she worked consistently over the day, she said, she could earn more than her polio pay by doing sewing piecework—and, she pointed out, she did not have to leave her children behind or walk around in the sun.

Overall, women said the challenges they faced in balancing home and work life would be manageable *if* they were making more money on polio days. Then, they could, for example, afford to give someone a bit of money to watch their children. Many women echoed the sentiment quoted above: "If the money was good, I could deal with the other issues."

The polio worker murders

In December 2012, nine polio workers, many of them female volunteers, were murdered as they delivered polio vaccines. Lethal attacks on polio workers continued into 2015, with more than 30 and perhaps as many as 70 workers murdered in total. These attacks were particularly shocking because there was little precedent for violence towards vaccination workers, even in Pakistan's many conflict zones. Several factors likely led to this outcome: the Central Intelligence Agency's (CIA's) use of a fake vaccination campaign in its search for Osama bin Laden; the high political profile of polio eradication; and the international shock value of attacking health workers (Abimbola et al. 2013; Closser and Jooma 2013). Despite the killings, Pakistan has not suspended polio campaigns—the danger of reintroducing polio to neighboring, polio-free countries like India and China makes stopping campaigns politically nearly impossible. But workers, understandably extremely concerned about the new lethal risks their job entails, have protested, asking for increased compensation and protection (Wasif 2012; Dawn.com 2013).

In the months after the initial killings, international agencies employing these workers used the language of heroism to describe the work done by their female volunteers. UNICEF and WHO (2012) released a statement saying: "Those killed or injured, many of whom are women, are among the hundreds of thousands of heroes who work selflessly to eradicate polio and provide other health services to children in Pakistan." This discourse of heroism leaves little room for discussion of labor practices.

Rethinking volunteerism

The Global Polio Eradication Initiative, then, like many other major global health projects, entrenches income inequality through its labor practices even as it lessens the impact of inequality in one very narrow realm, that of poliovirus transmission. It exploits youth labor in the service of child health. But these contradictory impacts of the Global Polio Eradication Initiative are in fact linked. The fact that the project aims to assist poor children is precisely what allows it to exploit poor adolescents.

In the international moral economy of humanitarianism and health aid, Didier Fassin (2013) has argued, children are viewed as "sinless and powerless" victims. In this moral economy, helping children is the greatest good. Tenuous labor practices become acceptable, even desired, in the pursuit of this larger goal. (If paying ground-level staff less than a living wage means more children can be reached with lifesaving interventions, it becomes a step that seems acceptable, even necessary.) The root of the issue behind the exploitation of youth labor within polio eradication, then, is not that the ideal of "volunteerism" is lost in cultural translation (as is sometimes suggested by international policymakers). Rather, it is that the moral economy of saving children silences critique and buttresses labor exploitation.

A better way forward is possible: international agencies that employ underpaid "development" workers can choose to prioritize providing them with a living wage. Instead of taking advantage of high unemployment to exploit staff, they could choose to have dignified labor be a central part of their approach. They do not do so because they do not need to: young women in poor countries across the world will take work, even dangerous work, for low or no pay, in the hope of some future benefit that may never arrive.

Female youth unemployment in Pakistan is a real, widespread, and largely invisible problem, hidden behind the ideal of women as non-workers. But it is this context of high unemployment that pushes young women into part-time and underpaid "volunteer" work. Moneyed global health programs allow conditions of high unemployment to drive workers towards below-minimum-wage jobs, and paint over this action with vaunted language about volunteerism. In the mean-time, the families of these underpaid workers struggle to put food on the table.

Notes

1 Much of this work, say the researchers, went unreported when respondents were asked if they worked for pay, but was picked up in time-use profiles.

2 The close ties of *purdah* to ethnicity and class, in addition to religion, explain why women in other Muslim countries like Bangladesh and Indonesia participate in formal labor markets like garment work in ways that are largely closed to Pakistani women.
3 Afghanistan had 28, Nigeria (which had not yet eliminated polio) had 36, and the rest were smaller, shorter-lived outbreaks in various parts of the globe.
4 Because donor money may not have been forthcoming for the full amount, actual expenditure may have been somewhat less.
5 Workers received a raise of 100 rupees per day in late 2011; however, inflation subsequently eliminated that raise in real terms, so the functional pay of polio workers in 2014 was again the same as it was when these interviews were carried out in mid-2011.

References

Abimbola, Seye, Asmat Ullah Malik, and Ghulam Farooq Mansoor. 2013. "The Final Push for Polio Eradication: Addressing the Challenge of Violence in Afghanistan, Pakistan, and Nigeria." *PLoS Medicine*, 10(10): e1001529.

Akhtar, Aasim Sajjad. 2013. "Dependency Is Dead: Long Live Dependency." In *Development Challenges Confronting Pakistan*. A.M. Weiss and S.G. Khattak. eds Sterling, VA: Kumarian Press, pp. 43–56.

Akintola, Olagoke. 2008. "Unpaid HIV/AIDS-Care in Southern Africa: Forms, Context, and Implications." *Feminist Economics*, 14(4): 117.

Aylward,Raymond Bruce, Harry Frazer Hull, S.L. Cochi et al. 2000. "Disease Eradication as a Public Health Strategy: A Case Study of Poliomyelitis Eradication." *Bulletin of the World Health Organization*, 78(3): 285–297.

Becker, Felicitas. 2015. "Obscuring and Revealing: Muslim Engagement with Volunteering and the Aid Sector in Tanzania." *African Studies Review*, 58(2): 111–133.

Biehl, João. 2007. *Will to Live: AIDS Therapies and the Politics of Survival*. Princeton, NJ: Princeton University Press.

Brickell, Katherine, and Sylvia Chant. 2010. "'The Unbearable Heaviness of Being': Reflections on Female Altruism in Cambodia, Philippines, the Gambia, and Costa Rica." *Progress in Development Studies*, 10(2): 145–159.

Brown, Hannah, and Ruth J. Prince. 2015. "Introduction: Volunteer Labor—Pasts and Futures of Work, Development, and Citizenship in East Africa." *African Studies Review*, 58(2): 29–42.

Chant, Sylvia. 2008. "The 'Feminisation of Poverty' and the 'Feminisation' of Anti-Poverty Programmes: Room for Revision?" *Journal of Development Studies*, 44(2): 165–197.

Chant, Sylvia. 2016. "Women, Girls and World Poverty: Empowerment, Equality or Essentialism?" *International Development Planning Review*, 38(1): 1–24.

Closser, Svea. 2010. *Chasing Polio in Pakistan: Why the World's Largest Public Health Initiative May Fail*. Nashville, TN: Vanderbilt University Press.

Closser, Svea, and Rashid Jooma. 2013. "Why We Must Provide Better Support for Pakistan's Female Frontline Health Workers." *PLoS Medicine*, 10(10): e1001528.

Closser, Svea, Kelly Cox, Thomas M. Parris et al. 2014. "The Impact of Polio Eradication on Routine Immunization and Primary Health Care: A Mixed-Methods Study." *Journal of Infectious Diseases*, 10: S504–S513. DOI: 10.1093/infdis/jit232.

Closser, Svea, Rashid Jooma, Emma Varley et al. 2015. "Polio Eradication and Health Systems in Karachi: Vaccine Refusals in Context." *Global Health Communication*, 1(1): 32–20.

Daniel, Thomas M., and Frederick C. Robbins. 1999. *Polio*. Rochester, NY: University Rochester Press.

Dawn.com. 2013. "Lady Health Workers in a Fix about Threats." Retrieved from http://dawn.com/2013/01/08/lady-health-workers-in-a-fix-about-threats (accessed February 8, 2013).

Drager, Sigrid, Gulin Gedik, and Mario R. Dal Poz. 2006. "Health Workforce Issues and the Global Fund to Fight AIDS, Tuberculosis, and Malaria: An Analytical Review." *Human Resources for Health*, 4(1): 23–24.

Dworkin, Barry. 2007. "Update on the Global Polio Eradication Program of the World Health Organization," July 8. Retrieved from www.drbarrydworkin.com/2007/07/08/update-on-the-global-polio-eradication-program-of-the-world-health-organization (accessed August 8, 2014).

Fassin, Didier. 2007. "Humanitarianism: A Nongovernmental Government." In *Nongovernmental Politics*. M. Feher, ed. New York: Zone Books, pp. 149–160.

Fassin, Didier. 2013. "Children as Victims: The Moral Economy of Childhood in the Times of AIDS." In *When People Come First*. João Biehl and Adriana Petryna, eds. Princeton, NJ: Princeton University Press, pp. 109–130.

Glenton, Claire, Inger B. Scheel, Sabina Pradhan et al. 2010. "The Female Community Health Volunteer Programme in Nepal: Decision Makers' Perceptions of Volunteerism, Payment and Other Incentives." *Social Science & Medicine*, 70(12): 1920–1927.

Global Polio Eradication Initiative. 2015. *Pakistan: Details of External Funding Requirements for 2014–2015*. Geneva: World Health Organization. Retrieved from www.polioeradication.org/Portals/0/Document/Financing/Onepager_Pak.pdf

Gold, Ann. 1994. "Purdah is as Purdah's Kept." In *Listen to the Heron's Words*. G. Raheja and A. Gold. eds Berkeley, CA: University of California Press, pp. 164–181.

Government of Pakistan Ministry of Labour and Manpower. 2009. *Pakistan Employment Trends for Women*. Islamabad: Government of Pakistan.

Government of Sindh. 2013. *Sindh Employment Trends 2013: Skills*. Karachi: Government of Sindh.

Independent Monitoring Board of the Global Polio Eradication Initiative. 2013. *Seventh Report*. London: Independent Monitoring Board of the Global Polio Eradication Initiative.

International Labour Organization (ILO). 2015. "Decent Work Agenda." Retrieved from www.ilo.org/global/about-the-ilo/decent-work-agenda/lang–en/index.htm (accessed July 20, 2015).

Jalal, Ayesha. 1991. "The Convenience of Subservience: Women and the State of Pakistan." In *Women, Islam, and the State*. D. Kandiyoti, ed. Philadelphia, PA: Temple University Press, pp. 77–114.

Kalofonos, Ippolytos. 2014. "'All They Do Is Pray': Community Labour and the Narrowing of 'Care' during Mozambique's HIV Scale-Up." *Global Public Health*, 9(1–2): 7–24.

Kazi, Shahnaz, and Bilquees Raza. 1989. "Women in the Informal Sector: Home-Based Workers in Karachi." *The Pakistan Development Review*, 28(4 Part II): 777–788.

Khan, Ayesha. 2007. *Women and Paid Work in Pakistan*. Karachi: Collective for Social Science Research.

Khan, Ayesha. 2011. "Lady Health Workers and Social Change in Pakistan." *Economic and Political Weekly*, 46(30): 28–31.

Lloyd, Cynthia, and Monica Grant. 2004. *Growing Up in Pakistan: The Separate Experiences of Males and Females*. Islamabad: Population Council.

Maes, Kenneth. 2010. "Examining Health-Care Volunteerism in a Food- and Financially-Insecure World." *Bulletin of the World Health Organization*, 88(11): 867–869.

Maes, Kenneth. 2012. "Volunteerism or Labor Exploitation? Harnessing the Volunteer Spirit to Sustain AIDS Treatment Programs in Urban Ethiopia." *Human Organization*, 71(1): 54–64.

Maes, Kenneth C., Brandon A. Kohrt, and Svea Closser. 2010. "Culture, Status and Context in Community Health Worker Pay: Pitfalls and Opportunities for Policy Research. A Commentary on Glenton et al. (2010)." *Social Science & Medicine*, 71 (8): 1375–1378.

Makino, Momoe. 2012. "What Motivates Female Operators to Enter the Garment Industry in Pakistan in the Post-MFA Period?" IDE Discussion Paper No. 374. Chiba, Japan: Institute of Developing Economies.

Mies, Maria. 1981. "Dynamics of Sexual Division of Labour and Capital Accumulation: Women Lace Workers of Narsapur." *Economic and Political Weekly*, 16(10–12): 487–500.

Mumtaz, Zubia, Sarah Salway, Muneeba Waseem, and Nighat Umer. 2003. "Gender-Based Barriers to Primary Health Care Provision in Pakistan: The Experience of Female Providers." *Health Policy and Planning*, 18(3): 261–269.

Mumtaz, Zubia, Sarah Salway, Candace Nykiforuk et al. 2013. "The Role of Social Geography on Lady Health Workers' Mobility and Effectiveness in Pakistan." *Social Science & Medicine*, 91: 48–57.

Ooms, G., Wim Van Damme, and Marleen Temmerman. 2007. "Medicines without Doctors: Why the Global Fund Must Fund Salaries of Health Workers to Expand AIDS Treatment." *PLOS Medicine*, 4(4): 605–608.

Papanek, Hannah. 1973. "Purdah: Separate Worlds and Symbolic Shelters." *Comparative Studies in Society and History*, 15(3): 289–325.

Ramirez-Valles, Jesus. 1998. "Promoting Health, Promoting Women: The Construction of Female and Professional Identities in the Discourse of Community Health Workers." *Social Science & Medicine*, 47(11): 1749–1762.

Ravishankar, Nirmala, Paul Gubbins, Rebecca J. Cooley et al. 2009. "Financing of Global Health: Tracking Development Assistance for Health from 1990 to 2007." *The Lancet*, 373(9681): 2113–2124.

Redfield, Peter. 2006. "A Less Modest Witness: Collective Advocacy and Motivated Truth in a Medical Humanitarian Movement." *American Ethnologist*, 33(1): 3–26.

Sassen, Saskia. 2000. "The Global City: Strategic Site/New Frontier." *American Studies*, 41(2–3): 79–95.

State Bank of Pakistan Statistics and DWH Department. 2011. *Inflation Monitor July 2011*. Islamabad: Government of Pakistan.

Swidler, Ann, and Susan Cotts Watkins. 2009. "'Teach a Man to Fish': The Sustainability Doctrine and Its Social Consequences." *World Development*, 37(7): 1182–1196.

Thigpen, Scott (director). 2004. *The Last Child: The Global Race to End Polio*. Bullfrog Films.

Tichtin, M. 2006. "Where Ethics and Politics Meet: The Violence of Humanitarianism in France." *American Ethnologist*, 33(1): 33–49.

United Nations Children's Fund (UNICEF) and World Health Organization (WHO). 2012. "UNICEF and WHO: Polio Vaccinators Attacked in Pakistan Are 'Heroes.'" UNICEF USA. Retrieved from www.unicefusa.org/press/releases/unicef-and-who-polio-vaccinators-attacked-pakistan-are-%E2%80%9Cheroes%E2%80%9D/8190 (accessed August 15, 2014).

Wasif, Sehrish. 2012. "Polio Campaign: Distressed with Security and Monetary Concerns, LHWs Camp Outside PM Secretariat." *The Express Tribune*, December 23. Retrieved from http://tribune.com.pk/story/483148/polio-campaign-distressed-with-security-and-monetary-concerns-lhws-camp-outside-pm-secretariat (accessed February 11, 2013).

Weiss, Anita M. 1992. *Walls within Walls: Life Histories of Working Women in the Old City of Lahore*. Boulder, CO: Westview Press.

World Bank. 2014. "Data: Pakistan." Retrieved from http://data.worldbank.org/country/pakistan#cp_wdi (accessed August 15, 2014).

World Health Organization (WHO). 2003. *Global Polio Eradication Initiative Strategic Plan 2004–2008*. Geneva: WHO.

World Health Organization (WHO). 2014a. "Countries: Pakistan." WHO. Retrieved from www.who.int/countries/pak/en (accessed August 15, 2014).

World Health Organization (WHO). 2014b. *Global Polio Eradication Initiative Financial Resource Requirements 2013–2018 as of 1 February 2014*. Geneva: WHO.

Zaidi, S. Akbar. 2011. "Who Benefits from US Aid to Pakistan?" *Economic & Political Weekly*, August 6. Retrieved from www.lexisnexis.com.ezproxy.middlebury.edu/lna cui2api/api/version1/getDocCui?lni=53HG-CBR1-JB35-10HW&csi=365197&hl=t&hv=t&hnsd=f&hns=t&hgn=tô00240&perma=true

5 Dealing with joblessness

Young people's life trajectory through "non-work" activities in Buenos Aires, Argentina

Mariano D. Perelman

For decades, the labor market in Argentina was configured as the legitimate way to move towards social reproduction. Work functioned not only as a natural path in the transition from youth to adulthood, but also in many cases as the moral vector of social behavior. Since the last civic-military dictatorship (1976–1983), however—which through repressive and bloody action began implementing neoliberal policies that were taken to their maximum expression during the governments of Carlos Menem (1989–1999)—the structure of the labor market changed. The neoliberal era eroded and transformed the social construction of work in different ways depending on the past experience of family, work, and lived experience, and on the age of the "workers." If in this process moral assessments regarding work have been questioned (Perelman 2014), these differences are even stronger in relation to young people, especially children (Llobet 2012). This chapter focuses on societal change since the implementation of a series of neoliberal policies in Argentina, and it examines their impact on the subjectivities of young people in two occupations designated as non-work activities: waste picking and street vending. This chapter contributes to an insufficiently explored area in the study of youth joblessness; by analyzing data collected through fieldwork it reveals the way in which flesh-and-blood young people have lived through these processes, and how in a context of labor precarity the traditional limits between being employed and unemployed acquire new edges.

Facing the options that emerged for these young people, waste picking for some and street vending for others were configured as ways for them to access a sustainable life. I will focus on the young people who perform these activities but have never held a formal job. (In many cases their parents have never held a stable job either.) Waste collecting and street vending cannot be said to "have a starting point" for young people any more than these young people's work trajectories can be seen as independent from their life trajectories. The young people I have conducted fieldwork with, many of whom were born during the 1980s and 1990s, have lived through structural economic transformations and financial crises that have had an impact on them and the conditions in their homes. These young people experienced a relationship to work marked by labor rotation, temporary jobs, and social protection benefits (made flexible, unstable, or lacking), all linked to precarious economic conditions. But young people's immersion in the stigmatized activities of waste picking and street vending can only be understood in relation to moral valorizations and the ways in which these types of work have become naturalized in day-to-day life through family experience.

Following other studies (Perelman 2011a; de L'Estoile 2014; Narotzky and Besnier 2014), I am concerned with the way in which ordinary people understand the concept of "life worth living" and what they do to strive towards that goal. In that sense, as Narotzky and Besnier (2014: S5) have pointed out: "Social reproduction entails addressing different scales in terms of which ordinary people evaluate the possibility of continuities, transformations, or blockages." My aim is not to discuss "informal garbage collection" or ambulant vending as a global activity;[1] this is the way in which these activities in other parts of the world are often referenced. Instead, my point is to demonstrate that the meaning of being a waste collector or street vendor in Buenos Aires depends on understanding its context within Argentinean history and the peoples' trajectory, and how social problems are historically constructed.[2] The analysis of these two activities will provide insights into the way young people perceive their possibilities in life, and the way moral choices are made in a context of great inequality and exploitation. The first section explores transformations of the labor market and the second section explains the insertion of young people in the aforementioned activities.

Neoliberal transformations

Before the *coup d'état* of March 1976 that ushered in seven years of violent military rule, Argentina was an almost full-employment society. Argentines understood their participation in the formal workplace—often by way of the Peronist state and its successors—as a legitimate way to consume, and to reproduce life models (Grassi 2003). If until the 1960s there was a period of high employment growth and socioeconomic improvements (Lindenboim 2008), after the military coup in 1976 the reverse was the case, and conditions worsened at an accelerated pace during the long neoliberal era (1989–2002). From 1975 to 2001, the Permanent Household Surveys run by the National Institute of Statistics and Census[3] show an unemployment rise in Greater Buenos Aires from 2.4% in April 1975 to 17.4% in 2001. Underemployment went from 4.7% in 1975 to 15.6% in 2001. Following the economic turmoil at the end of President Raúl Alfonsín's government (1983–1989), Menem's governments (1989–1999) generated the idea that the neoliberal model was the only solution to the social crisis. Under Menem, the roles of the state and trade unions were radically reduced. State companies and providers of basic services, including those for natural gas, water, telephone, railway transport, airlines, and oil, were privatized in giveaways to foreign and domestic investors. Thousands lost their jobs in the deindustrialization of the economy that ensued.

The establishment of a one-to-one parity between the Argentine peso and the US dollar in 1991,[4] as labor legislation and social policy, discarded the Peronist era notion of heightened expectations for workers in favor of bare minima for survival (Basualdo 2001). As the government enacted neoliberal precepts of structural adjustment in the construction of its neoliberal state, finance capital took the place of social policies. These neoliberal transformations have had an undeniable effect on the ways in which Argentines imagine what constitutes legitimate work. In conjunction with growing unemployment, the dispute over what work is and what it is not expanded and crossed new borders. There was a strong erosion of the social citizenship model associated with the welfare state of the mid-20th century. This process has had a significant impact on the country's young people. Unemployment numbers show these differences. According to the Argentine Ministry of Work, young people are three times more likely to be unemployed than adults. Moreover, young people suffer a high degree of rotation and greater circulation between activity and inactivity and, within the workforce, between employment, unemployment, and precarious jobs (Maurizio 2011). This is reflected in the unemployment rates of the 1990s.

I do not address the macroeconomic transformations expressed in the data (these are well accounted for by statistical research on youth work trajectories in Jacinto and Chitarroni 2011). Rather, I focus on the life experiences of a group of poor young people who, in spite of increasing difficulties, sought to earn a living by working. In this way, as noted by Chaves (2007: 4), although there have been numerous investigations on the young people of "popular" sectors—and there is information on the economic condition of young people mainly constructed by statistics and in relation to the formal and informal labor markets—"qualitative studies addressing the situation and work trajectories of youth are still insufficient" (Chaves 2007: 6). Thus, Chaves (2007: 6) points to the urgent need for studies on the work history of young people, not only studies that would account for "the significant relationship young people establish (or don't) with work and/ or employment, but also research on young people's perceptions of their future work lives, including youth's valorization of work." To make such research feasible, rather than just focusing on statistical data for work trajectories, it is necessary to reconstruct a diverse system of social relations and to account for a representational system linked to work that undoubtedly changed during the neoliberal decade, but still permeates young people's social imaginary as the way to legitimately access social reproduction.

In relation to the social, economic, and political changes that occurred during the 1990s, Merklen (2000) referred to people living in a situation of vulnerability as following a hunting (of resources) logic. In contrast to the stability and integration previously provided by work (and education),[5] people living in settlements in Greater Buenos Aires (where many of the young waste pickers I have worked with come from) had to get by on a daily basis in order to survive. According to Merklen (2000), these people experienced a feeling of estrangement in this process of social disaffiliation and downward social mobility.

Similarly, Kessler's (2013) study on crime based on juvenile offenders differentiated three moments in the relationship between labor markets and crime: the first one between the 1960s and the 1980s, the second comprising the 1990s, and the third one beginning in 2003. Significantly, his historical analysis enabled an understanding of some of the young people's more general problems and their relationship to work (both as a source of income and as moral value). According to Kessler (2013: 130), the 1990s marked the emergence among young people of a provision rationale based on theft, combining legal and illegal aspects. The author pointed to a transition from a worker rationale, predominant in a previous era, to a provider rationale. These young people are the children of already economically unstable parents. The difference lies in the legitimacy of the obtained resources. In the worker rationale, legitimacy lies in the origin of money (as the result of honest work in a respectable and socially recognized occupation), while in the provision rationale legitimacy no longer lies in the money's *origin*, but in its use to satisfy needs: any resource is legitimate if it enables covering a need (one that is lacking either totally or relatively). Young people who had turned to street vending or to waste collecting, just like the juvenile offenders Kessler examined, have parents who entered the labor market during the 1980s and 1990s in unstable terms. In contrast to the young people committing crimes, however, they had chosen from among the possible repertoire of options a rationale of obtaining rather than one of provision. They continued marking the origin of money (i.e., the way of obtaining money) as a legitimate way of making a living. It is a work rationale, albeit a new one generated in the light of neoliberal transformations.

Assigning terms in context: cirujas *and* cartoneras

Cirujeo is the activity of informal waste collecting. The term *ciruja*—as the people that perform *cirujeo* are known—entails a transformation of the word *cirujano* (surgeon) and can also be translated as "scavenger" or "picker." Despite the different forms this activity has taken and the different names assigned to the people carrying it out, *cirujeo* and its practitioners (throughout a long history of transformations) have always been stigmatized and have always been linked to both poverty and marginalization (Perelman 2012). As I point out elsewhere (Perelman 2016), in contrast to the subjects of other studies on unemployed populations who once considered themselves "stable" workers or who used to have stable jobs, many *cirujas* never knew stability in this way. For many people who were to become waste pickers, instability was the norm and they learned to live with it. But this instability became increasingly more pressing after the mid-1990s, when unemployment soared to historic levels (see Figure 5.1 and Table 5.1). In a context where employment was scarce and entire neighborhoods were forced to do anything to survive, in times when social assistance was insufficient, "looking for a job" was useless. Therefore, thousands of people began to look for alternative ways of living. This process, involving entire families, unfolded during the 1990s and the first years of the 21st century.

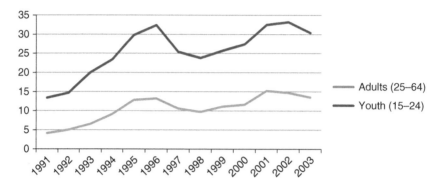

Figure 5.1 Youth and adult unemployment (1991–2003)

Table 5.1 Youth unemployment (1991–2003)

Year	Adults (25–64)	Young People (15–24)
1991	4.1	13.4
1992	5.0	14.7
1993	6.5	20.0
1994	9.1	23.4
1995	12.8	29.8
1996	13.2	32.4
1997	10.6	25.4
1998	9.7	23.8
1999	11.1	25.7
2000	11.6	27.4
2001	15.2	32.5
2002	14.7	33.2
2003	13.5	30.4

Source: Author's own elaboration based on Instituto Nacional de Estadística y Censos (INDEC) data.

In this new experience it is possible to see some common processes, ways of acting and feeling, regarding (the lack of) work. Even many *cirujas* who never had a formal job compared their situation to such jobs or to a time when stability was the rule. Many *cirujas* wanted to look for other forms of work, but they did not. Many *cirujas* were ashamed of waste collection, but they were also proud of it. Seeing themselves as workers, but at the same time as being unemployed, provided the backdrop to these feelings and transformations, and to the way in which the activity of waste picking itself was performed. These simultaneously experienced feelings of shame and pride can only be understood in the context of Argentine history. On the one hand, the archetype of the worker seeking a legitimate way of earning a living would never see informal waste collection as an option, hence the shame attached to the activity. Pride, in turn, is constructed in opposition to the idea of theft and in relation to the establishment of the notion of *cirujeo* as work. This process is significant because, until the end of 2002, informal waste collection was prohibited; it was illegal and was considered to be "not a job" (Perelman 2011a). In this context, two processes linked to the entry of young people into *cirujeo* took place. Waste collecting had already become part of many young people's lives because they were already participating in the activity with their parents. Although most of the young people with whom I conducted fieldwork experienced entering *cirujeo* as a disruption in regard to work expectations, it was a more natural process than it was for their parents. The young people I focused on grew up in families that collected waste, naturalizing this work as well as the conditions that made getting a job impossible. In fact, many *cartoneros* (informal waste collectors) start their working lives during their youth. A 2006 report by the United Nations Children's Fund (UNICEF) and the International Organization for Migration (OIM) shows that almost half of all *cartoneros* in the autonomous city of Buenos Aires were boys, girls, and teenagers (UNICEF-OIM 2006: 9).

During my first years of fieldwork with *cartoneros* (which I began in mid-2002), hundreds of children and young people performed waste collecting with their parents or on their own. Those were times of "crisis" in which unemployment and poverty reached levels never formerly seen. Most of the young people had never worked or had held only occasional "odd jobs." The older ones had held precarious and short-term jobs, which progressively became even more precarious and of shorter duration. These young people's relationship to waste collection changed according to how long they had been performing the task, whether they had their own family, and in relation to gender and age. For the youngest ones, there seemed to be a ludic quality to *cirujeo*. Going out on the street was almost a game, which involved begging for change and searching for recyclable material within garbage bags; going out and being on the street was an experience that was both suffered and enjoyed. Children talked about going through garbage bags as if they were participating in a treasure hunt.

Young people's situations are a bit different because they are usually in charge of their own family or actively contribute to that of their parents. Young people, especially young women, constitute a much more vulnerable group than adults. Teen pregnancies are usually ubiquitous. During my fieldwork, many young women got pregnant. This did not stop them from going out to collect waste. In the same way as with children, naturalization operated here. In contrast to their parents, many of them did not "know" other ways of inserting themselves into the labor market. In 2007, Yanina, a 14-year-old girl, told me that:

> The family must eat, and my parents couldn't get a job. My father was depressed because he couldn't find a job and started collecting. I can't remember it but he told me he worked in a restaurant. For me it is funny to go to the city.

Frequently the question about how these young people started to collect or why they carried out this activity pointed to circumstance rather than preference: "I go out to work to help my family," 16-year-old Santino told me; or, "I've been collecting since I'm very young, I started out with my mother when she was left unemployed. After that I started going out on my own with my cart to earn bread for my family," said 17-year-old Nicolas. In contrast to the accounts of adults (Perelman 2011a), young people did not see their entry to *cirujeo* as a disruption in their trajectory or consider it a milestone activity in terms of pursuing a livelihood.

On one occasion, Marcos, a 14-year old, described how he and his father started collecting waste. During the 1970s his father (at the age of 17) had migrated to Buenos Aires from the province of Misiones, driven by his older brother who had promised him a job. That same year he started to work in a shipyard. Later on, he entered the army and after that he worked in private security. After losing his job in 1999, he was only able to obtain very sporadic work in construction. Without being able to carry out any activity that could ensure survival, and in times in which his whole family was also left unemployed, he began turning to *cirujeo*. I vividly recall Marcos's account:

> We had gone nine days without eating, I was out of strength. There was an abandoned car over where we lived. We went looking for something to take out of it. When we were crossing over a ditch my dad fell down, he was out of strength. I asked him to get up. We reached the car and were able to take some stuff out. That day we ate plain boiled rice; I'll never forget it. In those days we saw lines of people with carts but we didn't know where they were going. "We have to follow them," he told me [...] A few days later, when we started coming in [to the city of Buenos Aires], I went into shops to ask for pastries, bread, food. My old man didn't like it; he stood aside, half hidden.

Without questioning the truthfulness of the account about the length of time they went without eating, I am interested in highlighting the extent to which Marcos constructed the notion of how difficult it was for his father to decide to begin waste collecting. In contrast to Marcos's memories of his father's shame, for him not only *cirujeo* but also asking for food was a natural way of making a living.

Just like other waste collectors, Marcos differentiated *cirujeo* from theft, marking the distinction in the money's origin. "Here many of the kids steal from their neighbors" and "every day I go out with my cart in order to eat, I don't go stealing" were phrases I regularly heard. Discursively, these two ways of obtaining money were differentiated. The idea that work was the legitimate path for making a living continued to take precedence. And while for the parents—as well as for a large part of society—*cirujeo* had been left out as an option in the world of work, for young people this naturalization produced the notion of waste collecting as a legitimate job.

Street vendors

My first contact with young street vendors on trains took place by chance in one of my first approaches to the field. One afternoon while I was waiting for the train, a young man selling candy sat down next to me. I quickly introduced myself and told him I was doing research on people making a living on the train. Being young and distrustful of a stranger asking questions, he then tried to avoid me. But his attempt to do so provided direction to my research. Among other things, he said he "didn't know much about what I was asking him" because he had only been peddling for three months. He mentioned that he "came here because [he] was unemployed" and that nobody could hire him because he hadn't finished high school. But his account of street vending was ambivalent: it was a job, but he also did it because "nobody was going to hire him." I was perplexed for several reasons. I had recently conducted fieldwork with *cirujas* for whom, as I have already illustrated, the transition to the activity of waste collecting took place as a continuity of survival. My young interviewee, on the other hand, who was only 17, also referred to a lack of options: nobody was going to offer him a job, so he had decided to go out and sell candy. This conversation alerted me to the way in which he internalized the idea that "education" was a credential for getting a job. In contrast to the previous cases I had encountered, it seemed like he had to go out to work once he had completed his obligation of finishing school.

The context in which I conducted fieldwork with street vendors was different than with waste pickers. I began fieldwork with waste pickers almost a decade later, when the unemployment rate had changed: in 2002 it was about 20% and, in 2011, around 7%. Undoubtedly the Argentine context, family trajectories, and the presence of an ethnographer (as a receiver of a discourse), were important in the construction of that argument. Being unemployed in 2002 was not the same as in 2012, and my young interviewee knew it. Although I had previously experienced this situation with adults looking for work, I had not encountered it with *cirujas*. At least initially, this difference had to do with an issue similar to the one presented in the previous section: when I started looking into street vendors' trajectories I found that they had usually entered into the activity through relatives or acquaintances. My young interviewee started street vending sponsored by his maternal uncle, who sold notebooks on the same train; for him it was a highly sought-after activity.

It was possible for virtually anybody to "enter" the activity of waste picking; however, if you wanted to sell something on trains it was necessary to win your place. Young people starting out as vendors on trains did so either because another vendor backed them or by resorting to violence. The trajectories of many of these young street vendors was linked to street life and to a so-called marginal culture: drinking alcohol, violence, living on the streets, not having a job, or resorting to force for conflict resolution. The process of entering the activity took place during youth and was protected individually or, even more commonly, by a group.

Street vending is considered a legitimate way of earning a living and a step on the path of upwards social mobility. In contrast to *cirujeo*, which even when naturalized is collectively experienced as a sort of social downfall—as, for example, parents do not want their children to perform the task—street vending is subjectively experienced not only as a legitimate task, but as a desired one. People coming from families of street vendors teach their children or relatives and want them to sell; for people who live on the street it means leaving behind a universe of stigma and constant pursuit of resources for a much more stable existence.

For people who lived on the streets as children, street vending became a way of life in two respects: economically, it provided sufficient income to guarantee existence and, socially, it provided a sense of belonging. During my fieldwork I saw vendors getting together before starting the day, gathering to chat, to drink or use drugs, to divide the goods they were going to sell, and so forth. These relationships went far beyond the selling process itself. People who became street vendors became friends and shared a social life of which street vending was only a part.

Two of the closest friends I made during my fieldwork, Ramón (who is 41 years old) and Nestor (who is 43 years old), had similar trajectories. Ramón escaped from his home because his father beat him. He started living on the streets and selling things on trains. Once he told me: "Living in the street is not easy. I fought all days. I used drugs. But I started vending and I made friends and I started to be known by people. I feel great vending." Nestor started vending on the streets and on buses because his stepfather forced him:

> My father died and my mother married him because we had no money and he has a store. But he told me that he was not going to maintain me. I was 13. I spent most of the time in the street so I started meeting people that taught me how to sell. I am happy now of my work.

It is possible to find two different groups of young people vending on trains: those who come from families that have been vending for years and those who, due to long-term lack of employment and life on the streets, have found in this activity a way of making a living. A large part of the latter group started off as children begging for change on streets or trains, or stealing. Over time they found a place for themselves in street vending.

For these young men and women, both the rationale of origin and the rationale of utilization took precedence. Street vendors establish a double relationship with money: obtaining it and spending it in ways that produce group identification patterns through consumption. Although recent years have entailed economic growth in the labor market, instability for young people, especially the poor, emerges as an "implicit feature in all occupations, thus opportunities are seen as being short-term" (Kessler 2013: 159). In the accounts of street vending young people, in the same way as noted by Kessler, the provision rationale remains, but in the frame of a "democratization of consumption." During the Kirchner periods of government,[6] the experiences of impoverished sectors created new standards for assigning value and merit to ways of subsistence. If during the 1990s the meaning of work occupied a central place in subjectivity construction, the experiences of entire populations during neoliberalism debilitated the relationship between dignity and work.

With improvements in consumption, as well as the boost to consumption through improvements in objective life conditions, "strategies of distinction and valorization linked to certain goods become important, and a reconfiguration of relative deprivation takes place, to the extent that absolute deprivation has decreased" (Kessler 2013: 159). But neoliberal experience of deprivation is not only present in biographies, deposited under a layer of growth; these young people also live with the naturalization of violence and stigma attached to being poor and young. This has not only produced the legitimization of "alternative to work" forms, but also the naturalization of other practices (such as violence or drinking). In turn, this impacts the way of working or of looking for work. These young people "prefer" to sell on trains rather than to look for a job because they not only know (in terms of practical conscience) that it will be difficult or even impossible for them to get a job; they also accept their condition and, in a way, choose it.

Concluding remarks: youth unemployment and access to life

Let us return to some of the questions guiding this chapter as well as other work on unemployment: What do these cases tell us about unemployment and neoliberalism? What does it mean to be unemployed? Are there differences between youth and adult unemployment? Unemployment is a social category that enables the resignification of individual situations and differentiated social trajectories. As I point out, the analysis of real people in terms of their actions, choices, and practices provides a complex understanding of how people relate to work (as opposed to employment). Some social circumstances, such as the Argentinean crisis, changed the frame of experience and influenced the vision of what unemployment means and what defines a legitimated way of life. At the same time, trajectories constrain interpretations: everyone does not feel the same way about performing the same activity.

The category of unemployment expresses complex processes that imply a quantitative imbalance between the supply of and the demand for work. On the contrary, the unemployed are a socially constructed specific gender (Fernández Álvarez and Manzano 2007: 145–146). There is a normative and statistical understanding: some exchange relationships are considered employment, while others are not; however, this does not imply that people do not work. Another dimension, intimately related to the previous one, is the subjective process surrounding methods of social development. Here, this archetype of the worker, of someone "being employed," is embedded within social trajectories. In this sense, these young people, for different reasons, "know" that the formal labor market is not for them. Family trajectories show this. Hence, even if *cirujas* and street vendors work, "they are unemployed." In addition to not accessing social security benefits linked to employment (something they share with informal workers), they consider themselves unemployed and are regarded as such.

A way to address the specific differences or lack of them among young people carrying out these activities is to compare them with others who are not young. I suggest important differences exist that not only point to questions of age or circumstance, but also of experience. And by "experience" I refer to the trajectories of different generations that are making it possible today for young people to consider certain ways of life as natural.

As I mentioned above, poor young people are a stigmatized population and much more vulnerable than adults. That is, the objective conditions of a poor young man and a poor adult are different. This, of course, derives from their experience, and experience lays the groundwork for the ways of approaching life. The concept of "work" among adults—as a legitimate way of making a living in an era where it was possible to perceive a moral notion of work linked to one's welfare, effort, and decency—changed along with neoliberal policies that fostered individualism; however, work still remains as a moral value guiding actions and justifications. This "hegemonic" vision, however, is not homogeneous as Nestor's and Ramón's cases show. This is why frames of experiences and expectations are central to comprehending unemployment.

There are large differences within the age and cultural group called "youth." Transformations that occurred after the implementation of neoliberal policies (not only economic ones) have undoubtedly changed the ways of accessing social reproduction. Unemployment always serves as an element of discipline of the workforce. The fear of "losing a job" for example, has been an important element in the implementation of certain policies (those regarding the reduction of job benefits and the reneging of terms in contracts, for example), thus producing worse jobs, increasing poverty, and fostering the need to seek work at any cost. This fear is much more distant in young people than in adults because young people have become accustomed to living without a job. Neither their parents nor themselves, however, had experienced the result of contiguous decades of full employment and labor security. This is one of the main consequences of neoliberal policy: the naturalization of poverty and of precarious ways of living. The cases of *cirujeo* and street vending enable an understanding of the different ways in which social transformations accumulate, as a result of a network of horizontal and vertical relationships.

I started my work with young *cirujas* in the early 2000s, when Argentina was in the middle of a fierce economic crisis. My work with young street vendors, however, began in 2011, when unemployment and poverty rates had dropped by more than half. This is not the only information explaining significant differences between young *cartoneros* and street vendors; apart from showing the numbers of unemployed people, these rates also account for subjective experiences of social integration and the construction of legitimate ways to engage in life. In both cases, the deprivation is different. Waste collectors were driven by the need to survive. Young street vendors, however, although also motivated by the need to survive, live in a context that enables them to find collective practices of identification through consumption. In almost every case, these young men and women have families to support. Nevertheless, as they are not usually able to find a job (or get a desired job), they look for alternative ways of obtaining money. If previously the limit between youth and adulthood was marked by the transition into working activity—which does not imply that child and youth labor do not exist—currently these limits are blurred.

Cirujeo is a much more stigmatized activity than street vending. In this sense, if the notion of work appeals to young *cartoneros* it is because their parents attempted to make their past more comfortable by intertwining meanings attached to waste picking and work. This has had a double effect on young people. On the one hand, their search for survival is predominantly guided by a focus on the *origin* of money. On the other, an undesired activity has become ever-present in their social culture. Many *cartoneros* begin their working lives as children. Upon reaching youth they have achieved prolonged and naturalized experiences in a previously stigmatized task. In any case, *cirujeo* never appears as a desired work goal. Street vendors either come from families of vendors or come from a life on the streets, the difference being that the former usually finish their studies before they begin street vending, and the activity is a desired task. The latter, in contrast, start street vending at a younger age, and their continued engagement in it becomes a way of achieving upward mobility.

Many poor young men and women cannot find a place in the labor market due to objective conditions and entrenched social practices that affect people in real life. Currently, notwithstanding economic growth, access to the labor market is restricted. Young people are particularly vulnerable, but they also gradually create their own legitimate ways of making a living. This does not mean that unemployment is an individual problem (i.e., young people do not work because they do not want to). Cultural norms and lived social experiences of exploitation, as well as constant stigmatization, generate changing views of life in which practices that are alternative (and unequal) to the market are established today as naturalized activities.

Notes

1 There is a vast literature showing that "global" processes (such as neoliberalism) are locally experienced, appropriated, and answered (Edelman and Haugerud 2004; Ong 2006; Sharma and Gupta 2006; Gregory 2014). Along these lines, my chapter contributes to the way in which the processes of neoliberal social transformation had huge impacts on subjectivities.
2 It is more appropriate to compare ambulant vending or garbage collection with other forms of earning a living in Argentina than in other countries. See for a debate Perelman (2017).
3 See: www.indec.gov.ar
4 The National Law sanctioned in March 1991 established that, since April 1 of that year, the Argentinean national currency (the peso) and the US dollar were equivalents and interchangeable.
5 I go into this point only to highlight that, in spite of the author's valuable analysis, a close look at certain types of behavior would show that "hunters" are motivated by more than a mere search for opportunities. For the case of *cirujeo* see Perelman (2011b).

6 I refer to the governments of Néstor Kirchner (2003–2007) and Cristina Fernández de Kirchner (2007–2011 and 2011–2015). This stage entailed economic growth with notable social inclusion. In spite of this, the insertion of some groups is still shaky; however, an increase in social security (the broadening of pension plans and social compensations) has produced a noticeable improvement in the life conditions of poor sectors.

References

Basualdo, Eduardo. 2001. *Sistema político y modelo de Acumulación en la Argentina. Notas sobre el transformismo argentino durante la valorización financiera (1976–2001)* [*The Political System and Model of Accumulation in Argentina. Notes on the Argentine Transformation during the Financial Valorization (1976–2001)*]. Bernal: Universidad Nacional de Quilmes.

Chaves, Mariana. 2007. "Enfoques de las investigaciones de ciencias sociales sobre juventudes en Argentina" ["Approaches to Social Science Research on Youth in Argentina"]. In *Actas de la Primera Reunión Nacional de Investigadores/as en Juventudes*. La Plata: Universidad Nacional de La Plata, pp. 1–13.

Edelman, Marc, and Angelique Haugerud, eds. 2004. *The Anthropology of Development and Globalization: From Classical Political Economy to Contemporary Neoliberalism.* New York: Wiley-Blackwell.

Fernández Álvarez, María Inés, and Virginia Manzano. 2007. "Desempleo, acción estatal y movilización social en Argentina" ["Unemployment, State Action and Social Mobilization in Argentina"]. *Política y cultura*, 27: 143–166.

Grassi, Estela. 2003. *Políticas y problemas sociales en la sociedad neoliberal: la otra década infame* [*Policies and Social Problems in Neoliberal Society: The Other Infamous Decade*]. Buenos Aires: Espacio.

Gregory, Steven. 2014. *The Devil behind the Mirror: Globalization and Politics in the Dominican Republic* (1st edn with new preface). Berkeley, CA: University of California Press.

Jacinto, Claudia, and H. Chitarroni. 2011. "Precariedades, rotación y movilidades en las trayectorias laborales juveniles" ["Precariousness, Rotation and Mobility in Youth Work Trajectories"]. *Estudios del Trabajo*, 39–40: 5–36.

Kessler, Gabriel. 2013. "Ilegalismos en tres tiempos" ["Illegalities in Three Time Periods"]. In *Individuación, precariedad, inseguridad: "Desinstitucionalización del presente"* [*Individualization, Precarity, Insecurity: "Deinstitutionalization of the Present"*]. R. Castel, G. Kessler, D. Merklen, and N. Murard, eds. Buenos Aires: Paidós, pp. 109–166.

de L'Estoile, Benoit. 2014. "'Money Is Good, but a Friend Is Better': Uncertainty, Orientation to the Future, and the Economy." *Current Anthropology*, 55(S9): S62–S73.

Lindenboim, Javier. 2008. "Auge y declinación del trabajo y los ingresos en el siglo corto de la Argentina" ["Rise and Decline of Work and Incomes in the Short Century of Argentina"]. In *Trabajo, ingresos y políticas en Argentina: contribuciones para pensar el siglo XXI.* J. Lindenboim, ed. Ciudad de Buenos Aires: Eudeba, pp. 23–67.

Llobet, Valeria. 2012. "Una lectura sobre el trabajo infantil como objeto de estudio. A proposito del aporte de Viviana Zelizer" ["A Reading about Child Labor as an Object of Study. About the Contribution of Viviana Zelizer"]. *Desarrollo económico*, 52(206): 311–328.

Maurizio, Roxanna. 2011. *Trayectorias laborales de los jóvenes en Argentina: Dificultades en el mercado de trabajo o carrera laboral ascendente* [*Work Trajectories of Young*

People in Argentina: Difficulties in the Labor Market or in their Ascending Career]. Chile: CEPAL.

Merklen, Denis. 2000. "Vivir en los márgenes: la lógica del cazador. Notas sobre sociabilidad y cultura en los asentamientos del Gran Buenos Aires hacia fines de los 90" ["Living in the Margins: The Logic of the Hunter. Notes on Sociability and Culture in the Settlements of Greater Buenos Aires in the Late 1990s"]. In *Desde abajo: la transformación de las identidades sociales* (1st edn). M. Svampa, ed. Buenos Aires: Universidad Nacional de General Sarmiento, Editorial Biblos, pp. 81–119.

Narotzky, Susanna, and Niko Besnier. 2014. "Crisis, Value, and Hope: Rethinking the Economy: An Introduction to Supplement 9." *Current Anthropology*, 55(S9): S4–S16. doi: 10.1086/676327.

Ong, Aihwa. 2006. *Neoliberalism as Exception: Mutations in Citizenship and Sovereignty.* Durham, NC: Duke University Press.

Perelman, Mariano D. 2011a. "La construcción de la idea de trabajo digno en los cirujas de la ciudad de Buenos Aires" ["The Construction of the Idea of Decent Work in the Informal Garbage Collectors of the City of Buenos Aires"]. *Intersecciones en antropología*, 12(1): 69–81.

Perelman, Mariano D. 2011b. "La estabilización en el cirujeo de la ciudad de Buenos Aires. Una aproximación desde la antropología" ["The Stabilization of Informal Garbage Collection in the City of Buenos Aires. An Approach from Anthropology"]. *Desarrollo económico*, 51(201): 35–57.

Perelman, Mariano D. 2012. "Caracterizando la recolección informal en la ciudad de Buenos Aires" ["Characterizing Informal Garbage Collection in the City of Buenos Aires"]. *Latin American Research Review*, 47(Special Issue): 49–69.

Perelman, Mariano D. 2014. "Viviendo el trabajo. Transformaciones sociales, cirujeo y venta ambulante" ["Living the Work. Social Transformations, Informal Garbage Collection and Street Vending"]. *Trabajo y Sociedad*, 23: 45–65.

Perelman, Mariano D. 2016. "Contesting Unemployment: The Case of the *Cirujas* in Buenos Aires." In *Anthropologies of Unemployment: The Changing Study of Work and Its Absence.* J. Bum Kwon and C.M. Lane, eds. Ithaca, NY: Cornell University Press, pp. 97–117.

Perelman, Mariano D. 2017. "Collecte des déchets, crise et problèmes sociaux associés" [Collection, Crises and Social Problems behind Garbage]. In *Jeux de pouvoir dans nos poubelles. Enjeux idéologiques, sociaux et politiques du recyclage au tournant du 21e siècle.* É. Ansttet and N. Ortar, eds. Paris: Petra, pp. 161–178.

Perelman, Mariano D. In press. "Inequality and Marginalization. An Ethnography of Argentinean Urban Precarity." In *Routledge Handbook of Anthropology and the City. Engaging the Urban and the Future.* S. Low, ed. London: Routledge.

Sharma, Aradhana, and Akhil Gupta, eds. 2006. *The Anthropology of the State: A Reader* (1st edn). New York: Wiley-Blackwell.

United Nations Children's Fund and International Organization for Migration (UNICEF-OIM). 2006. *Informe sobre trabajo infantil en la recuperación y reciclaje de residuos* [*Report on Child Labor in the Recovery and Recycling of Waste*]. Buenos Aires: UNICEF-OIM.

6 Contrasting discourses surrounding gendered representations of young migrants negotiating for work on the South African border

Stanford T. Mahati

Introduction

The lived experiences of migrant children offer a useful vantage point from which to study youth unemployment in southern Africa. Of particular interest are the gendered expectations that shape local understanding of childhood and paid work. In this chapter I draw on ethnographic research conducted in the precarious environment of Musina, a town located on the Zimbabwe–South Africa border. Migrant children travel across that border, unaccompanied by parents or guardians, in their quest for jobs and improved lives in South Africa. This analysis examines the internal contradictions and ambiguities in the discourse of humanitarian aid workers who serve these children. Advancing Meyer's (2007: 87) view that "different social issues tend to be marked by the predominance of different discourses," I show how anti- and pro-child work discourses take shape, unfold, and function; I examine as well the consequences for young migrants faced with the challenge of unemployment in contexts where both discourses operate at different moments. In this chapter I suggest that certain dominant views either promote or serve to frustrate efforts to resolve the problem of youth unemployment. Intervention agencies need to be aware of these conflicting discourses and their consequences on young people who are trying to work their way out of poverty, especially those contending with migration and gendered workspaces.

This chapter follows and expands Nieuwenhuys's (1996: 238) call that future research should be "based on the idea of work as [one] of the most critical domains in which poor children can contest and negotiate childhood." In resource-poor settings where adults are unable to support their children, there is enormous pressure on young people to be employed full-time or temporarily in order to contribute to household economies (Bourdillon 2008). In turn, examining these pressures and practices fosters an understanding of how children are perceived to confirm or refute dominant notions of childhood and vulnerability (Bourdillon et al. 2010).

To distinguish between aid workers' formal and informal interactions, I define formal discourse to be not only what aid workers say in the public arena or in discussion with service providers and state officials, but also what government policies and organizational mission statements convey. Informal discourse, on the other hand, is what aid workers say outside those settings, which sometimes contradicts their organization's formal position. Arguably, formal and informal discourses define and structure "the ways in which the world, or parts of it, is to be understood and talked about" (Finnström 2006: 206).

Although there is abundant literature on the discourse concerning "child labor" and the larger phenomenon of children's work (Bourdillon et al. 2010), this chapter specifically focuses on an understanding of the "scaffolds of discursive frameworks" (Cheek 2004: 1142), by looking at working independent female migrant children within the context of a humanitarian crisis. In analyzing the representations of these children, I apply the point made by Cheek (2004: 1143): the particular discursive frame that is "afforded presence is a consequence of the effect of power relations." This position facilitates an understanding of just how many representations of independent female children looking for employment exist in the midst of aid workers' subjective relations with these children. Building on Meyer's (2007) insights about predominant discourses, I demonstrate the competing, contradictory discourses about childhood vulnerability that shape humanitarian workers' understanding of the young migrant girls they encounter, and how that in turn affects the unemployment of young migrants.

Contextualizing Zimbabwe and South Africa

The socioeconomic situation in Zimbabwe has a bearing on how independent migrant children from this country deal with the challenge of unemployment and how they are represented in South Africa. Zimbabwe's economy has been in a perilous state since the late 1980s, and it deteriorated further in the new millennium after the government embarked on its "fast track" land redistribution program (Chirisa and Muchini 2011; Raftopoulos 2009). The Economic Structural Adjustment Programme (ESAP) introduced in 1990 resulted in massive job losses, high levels of unemployment (including among young people), serious shortages of basic goods, the near collapse of the educational and health service delivery system, and a weak social protection system (Chirisa 2013; Mason 2009; Raftopoulos 2009). Consequently, children were increasingly drawn to work in the informal sector, particularly domestic and street work (Bourdillon 2009; Rurevo and Bourdillon 2003). As the Zimbabwean economy continued to deteriorate, a report produced by Malte Luebker in 2008 stated that youth unemployment accounted for 57.2% of the total unemployment (Luebker 2008). Many Zimbabwean young people abandoned their homes and left en masse for neighboring countries like South Africa—one of Africa's largest economies—in search of livelihoods and access to social services (see Rutherford 2008). If, according to Chirisa and Muchini (2011), the unemployment of young people is "a form of deprivation which robs youth of the benefits of work and represents a dark era in [their] personal and social development," then it is no wonder South Africa held the promise of a prosperous work life for the jobless and economically distressed Zimbabweans, despite its high levels of crime and violence, including xenophobic attacks (Landau 2010; Hassim et al. 2009). Much to the shock and distress of foreign migrants, however, South Africa is itself implementing neoliberal economic policies and has a high level of youth unemployment (Rankin and Roberts 2011) with around 48% of South Africans in the 15–34 age group being unemployed in the third quarter of 2016 (Statistics South Africa 2016).

The high-level presence of Zimbabwean migrants in South Africa, with many of them unemployed and struggling to get basic goods, generated a humanitarian crisis in Musina (Mahati 2015; Rutherford 2011). In response, different organizations set up services in the border town to assist foreign migrants (Mahati 2012a, 2012b). Between 2009 and 2010, I conducted ethnographic work in Musina, focusing on two organizations—one faith-based and the other an international humanitarian aid agency. In Musina, both agencies sought to care for and support independent young migrants they deemed vulnerable, physically weak, and incapable of supporting and protecting themselves against exploitative and dangerous work.

The group of about 20 aid workers I spoke with comprised Black South Africans and Zimbabweans, and non-African international workers. They did not consistently embrace the understanding of childhood put forth by either the South African government or their own organizations. Although the workers were mostly local, their official understandings drew from dominant global or Western ideologies about childhood and children's rights. In addition, these aid workers, the majority of whom were women, tapped into local ideologies of gender: that women should not seek employment but stay at home, and children who were away from home needed motherly care.

In the following sections I examine the two competing and not mutually exclusive discourses pertaining to child work: the official anti-child work approach and the unofficial pro-work approach, each of which mobilized at different moments to either legitimize or delegitimize not only young female migrants' attempts to be employed but also the practice of children working (Bourdillon et al. 2010; Mahati 2015; Twum-Danso 2013). I discuss how and why the pro-work approach often prevailed during unofficial encounters between aid workers and independent children. I contend that the contradictions in the discourses rest on social context, socioeconomic realities, different life-worlds, and understandings of childhood and girlhood.

The official discourse: anti-child work

South Africa is a signatory to the United Nations Convention on the Rights of Children (UNCRC), which tends to depict young people as vulnerable and in need of protection. In particular, within this framework, child labor is perceived as work that "impairs the health and development of children" (Fyfe, cited in James et al. 1998: 108). South African law prohibits children younger than age 15 from employment, and all children from dangerous and harmful work. Nevertheless, in South Africa, like many other African countries, work for children is perceived as part of the socialization practices inherent to childrearing. The current legal framework, however, views children as vulnerable. Contemporary laws have sought to criminalize the employment of children by making it difficult for children under age 18 to work legally or participate in certain work activities. This contradictory field of ideologies shapes aid workers' discomfort and anger towards the many young Zimbabwean migrants, including girls, working or looking for work.

In the discourse on children's rights (Ennew et al. 2005), formal interactions between migrant children and humanitarian workers tend to predominate, helping to delegitimize children's work and label working migrant children as deviants. This discourse, then, views the unemployment of young people (including migrant children) as problematic. In formal settings, most of the humanitarian workers' arguments highlight the negative effects of child labor, especially on young girls (Bourdillon et al. 2010). The effects of this discourse are compounded by some structural conditions. Lacking the work permits and tertiary qualifications that are prerequisites in South Africa's formal economy, a significant population of independent children are unemployed. A small population of young migrants who are working do so in the informal economic sector as vendors, domestics, and farm workers. It is often a herculean task for these young people, both living in and outside places of temporary safety, to get paid, negotiate for fair remuneration, and support their families. Meanwhile, drawing on their understandings of vulnerable, innocent childhood, humanitarian workers tend to pathologize working children and blame them for acting irresponsibly by appropriating adult activities. Further, existing data on children being abused and exploited at workplaces lend support to the discourse that children's agency at workplaces is limited.

The anti-child work discourse dominates official interactions between female working children and aid workers. Aid workers often tell migrant children to stop worrying about their unemployment status and stop assuming adult roles, and direct the girls among them to stop acting like boys. Humanitarian workers often emphasize that these children should be in school, not looking for employment. Thus, the workers are often at odds with migrant children who seek work or aim to both attend school and work.

By invoking the idea that childhood, particularly for girls, is a special state that needs protection, as well as "a special phase" (Clarke 2004: 9) that frees one from economic responsibility, these discourses emphasize that work (particularly monetized work) is not an arena for childhood and girlhood. As a result, working children, or children who are searching for jobs (especially girls) are seen as outliers who challenge the established social order that separates childhood and adulthood, boyhood and girlhood. This paternalism was articulated by both male and female humanitarian workers in my research, as well as by migrant boys; they all believed that girls should stay at home and avoid compromising or corrupting their womanhood. This official approach is influenced by both local and global understandings of childhood and the dominant discourse of the "girl-child being more vulnerable" and therefore in need of protection.

Humanitarian workers also perceived child work as interfering with the acquisition of education, and several were convinced that education would help to delay the girls' entry into marriages and childrearing. They often warned migrant girls that without an education they would be vulnerable to abuse by their future husbands and remain economic dependents. Aid workers often pointed out that lack of education causes unemployment and leads to ungainful employment. The position of these humanitarian workers reflects "the values and assumptions of a culturally bound and class-based discourse of childhood" and womanhood (Hoffman 2011: 2). The aid workers were drawing on Western discourses that view children as immature and mobile women as lacking sexual morals. Female migration was seen as violating appropriate gender norms, and independent migrant girls were seen as engaging in sex work and premarital sex. These results are consistent with other research (for example, Muzvidziwa 2001), indicating that there is a dominant discourse of associating independent female migrants with sexual immorality.

In the discourse of humanitarian workers, however, claims regarding the exploitation of migrant children were understood as a consequence of the children's naivety, not their poverty. Echoing Valentine's (2003: 37) point that "focusing on the North it is possible to argue that childhood is imagined as a time of innocence and freedom from responsibilities of adulthood," aid workers urged working children to abandon outside work or the attempt to find it. "Leave the responsibility of working to adults," said one aid worker. "What do you want to do when you grow up?" Arguably she was reinforcing middle-class Western notions that construct childhood as a period of dependency (see Orgocka 2012). Aid workers castigated work as irrelevant to young children's lives and also reinforced the idea that work is like "'an adults-only' site of knowledge, from which children, perceived to be too young to understand such knowledge, should be protected through the denial of access" (Robinson 2008: 121). Reinforcing "the idea of childhood as a special phase" (Clarke 2004: 9) in which "play" is the most appropriate activity for children (Jenks 1996), one aid worker chided working independent children who complained of exploitation on the job for having acted badly in the first place by taking on adult responsibilities. This victim-blaming approach often prevailed during informal interactions between aid workers and independent children. Aid workers often argued that children (particularly girls) were not physically fit to work or job hunt in high-risk spaces like private homes or on farms in South Africa, emphasizing that working is the domain of adults and men, and that migrant girls were behaving inappropriately by working. By so doing, aid workers effectively dismissed or marginalized the structural factors that force some children to seek employment. Aid workers' characterization of working children's childhood as "abnormal" (see Bourdillon 2006) was rooted in an understanding of work as a matter of choice, as freely available, and as depoliticized.

Since the physical and social cost of working for girls was seen as very high, aid workers also justified an anti-work sentiment by portraying work as an unprofitable venture for girls. Aid workers tended to expose and highlight the dangers that girls faced in the workplace, especially as domestic workers. Aid workers often cited harm to women's bodies, often in the form of sexual exploitation, to delegitimize the migrant girls' work or show the futility of girls' work in general, and to dissuade them from working and/or migrating. Drawing from the "children's best interest" idea—a children's rights principle stated in article 3 of the UNCRC—they normalized young migrants' unemployment much to the chagrin of young people who were hoping to make a difference to their lives by doing paid work.

The unofficial discourse: pro-child work

Anti-child labor discourse that has informed humanitarian agencies and government policies has tended to sideline concerns about why children have to be employed. While many aid workers formally approved of the governmental and non-governmental organizations' policies against child work, some social actors have challenged the anti-child labor discourse, particularly during informal situations. Arguably, this situation has emanated, according to Bourdillon and Spittler (2012: 11), from a local perspective: "The widespread view in African cultures is that work is essential to rearing children and preparing them for constructive adult life. According to this view, work provides necessary discipline and experience of responsibility." This mindset, which contradicts dominant discourses regarding the official interactions of children's rights and anti-child work, was widely shared by aid workers during informal situations. Drawing from the discourse of survival and children's responsibility to contribute to household economies in times of hardship, a number of aid workers backed children in their arguments that they had to do paid work to alleviate poverty in their lives as well as to support their families in Zimbabwe. One aid worker remarked: "I am impressed by their ability to save money and their unselfishness to use their money to buy basics for their siblings, parents, and even grandparents." Interestingly, these depictions of migrant children as responsible members of communities, with the financial competence to manage their wages, were not circulated during formal interactions as the aid workers opposed the official understandings of childhood and child work. In informal contexts, when children demonstrated an ability to save money or handle their earnings prudently, the discourses of childhood innocence, of girls being very vulnerable in workplaces, of children having freedom from economic responsibility, and of the need for children to be located in the domestic sphere, were silenced.

Considering that the major factor motivating migration is poverty—a condition that humanitarian agencies and other service providers do not have the resources to address—some aid workers questioned the rationale for stopping migrant children from looking for jobs and working, which allowed them to justify child work and oppose unemployment of poverty-stricken young migrants. Emphasizing children's poverty served to establish the innocence and victimhood of children looking for work on the move, contrary to the portrayal of them as deviants who have "self-destructive agency" (Gigengack 2008: 216). This showed the ambivalence of child labor discourse: recognized as bad on the one hand, yet seen as necessary in the lives of these poor children.

This acceptance of child work or employment of young migrants based on their obligation to support their families echoes the point Boyden (2009: 129) made in her study of how Ethiopian children contributed to their household's livelihoods: "coping with adversity is a collective rather than an individual responsibility." Actually, children in many societies assist their parents and guardians in raising their younger siblings (Lancy 2008). In the Musina study, aid workers also drew from this discourse by reminding children about their poverty-stricken family members and country, and by encouraging these children to send some remittances home. Aid workers often made disparaging remarks against migrant children they perceived as neglecting their responsibilities. Thus, contrary to the dominant discourse of employed girls being very vulnerable, they were not immune to these criticisms.

Despite the existence of anti-child work laws, several aid workers moralized child work. By doing so they rejected the discourses that framed childhood as fun, and as free from work and economic responsibility, and children as located solely in the domestic sphere. For instance, one of the aid workers argued that "the child labor law is not relevant to people who have left [behind] orphaned siblings to fend for." He agreed with aid workers who did not intervene against child workers younger than age 16 "as long as [the work] is not strenuous, exploitative, and in unfavorable working conditions." This reasoning buttressed arguments that children, including girls, can work or look for jobs as long as the work does not threaten their wellbeing.

Contradicting the official anti-child work position, aid workers tacitly tried to assist young migrants to deal with the problem of unemployment by linking them with potential employers, for example, and by giving them information on strategies and tactics regarding how and where to find employment. They supported and encouraged young migrants to work, helping them negotiate working conditions (in one case giving advice on getting time off to seek medical treatment), and sometimes assisting working young migrants engaged in labor disputes with their employers by safeguarding their earnings and groceries. Acknowledging the notion that children are "naïve and vulnerable" (Meyer 2007: 89), aid workers sometimes assisted these children by negotiating for fair employment remuneration and work conditions with their employers.

An unanticipated finding of this study was that, contrary to the official and dominant discourse of anti-child work, and in support of the idea that children should contribute to society if the situation demands, some aid workers celebrated working migrant children, both girls and boys, who were making a difference to their lives. Aid workers on the other hand had no kind words for unemployed young migrants they perceived as lazy (such as those who spent their days doing nothing at the shelter) and depended on aid. Aid workers' behavior served to embarrass but also teach and motivate other children to emulate the hard-working ones who often had money and regularly remitted money or groceries to their families (see Lancy 2008: 169–171). This situation further reinforced the point that the disapproval of children's employment was complex and situational.

The independent children (including those living in shelters) who continued to experience myriad problems like food shortages, and who felt they had a responsibility to assist their families who were living in abject poverty, have provoked researchers to wonder whose interests are being served when aid workers and organizations stop children from working or looking for jobs. "We cannot provide them with most of their needs and buy things like sweets for them," said a senior aid worker. In this construction, young migrants were expected and encouraged to partially fend for themselves. Explicitly or implicitly, aid workers were recognizing the potential of children (including girls) to take some control of their lives despite the many constraints they faced in this context (see Long 1992 on the actor-oriented approach and human agency). This indicates the limitations of the idealized global notion of childhood and girlhood, which views children regardless of gender as dependent and free from work with or without pay. Thus, the failure of aid workers to adequately provide young migrants in the child category with most of their needs forced them to "accept" or endorse child employment.

Consequences of the multiple and contradictory representations of independent migrant children

The different and contradictory representations of migrant children had different consequences. The dominant anti-child work discourse that associates child migration with anti-education, criminality, and exploitation leads to justification of practices like pathologization, unlawful detention, illegal deportation, police harassment of independent migrant girls, as well as clampdowns on the employment of these young people in South Africa. The anti-child work discourse did not allow for action like flexible schooling that accommodates child work, or for recognizing the role independent children play in supporting themselves and their families. In addition, these discourses did not allow support to be given to working young migrants or those actively looking for jobs.

Child protection was sometimes gendered and contradictory. Humanitarian workers depicted migrant girls as people who were "un-girl-like." This thinking allowed aid workers to contend that these children were responsible for any workplace problem in their lives even as they tried to deal with unemployment in a foreign country. Although humanitarian workers helped migrant girls particularly during formal situations, at times they withdrew the childhood status of these children, or denied their innocence based on a situational provision, just because they had crossed a border and were living and working on the streets.

Drawing from the discourse of formal schooling as the "rightful" activity for children, aid workers officially sought to keep these children on the periphery of the economy by supporting efforts to get them back to school either in South Africa or in Zimbabwe, and thus keeping them unemployed. Education was seen as very empowering to the girl child. As pointed out earlier, uneducated or less-educated girls were often projected as vulnerable even later in their lives. Thus the concerted efforts to send or keep children (especially girls) in school aimed to correct the situation and get migrant children to increase their marriage prospects, given the high value this society places on marriage. Single women were often viewed with scorn and associated with sexual immorality. Humanitarian workers (being adults and in *loco parentis*) said one of their major social responsibilities was not to assist these young migrants to tackle or reduce unemployment but instead to prepare girls for marriage and motherhood: teach them moral values like self-respect, self-discipline, patience, and caring.

Aid workers reinforced the notion of formal schooling as being good for children by either using subtle threats or making disparaging remarks about those children who were not attending school but instead concentrating on prospects for employment. As a result, some young migrant girls resorted to misrepresenting their intentions by pretending to focus on schooling when in fact they were actively looking for work and often engaged in temporary work. Fearing criticism from aid workers, they could not openly reveal their unemployment status. This situation, in which aid workers understood childhood as a time for learning (and for girls as a time to ready themselves for marriage), revealed that working migrant girls were alienated from the ideal state of childhood and girlhood. Working girls, or girls looking for jobs, and thus seen as undermining their chances of having a better life, were again portrayed as immature, further entrenching patriarchy—a position that functioned to silence these children's views and legitimize the imposition of adult views, as well as to prevent interventions aimed at empowering the girl child.

The expression of anti-child work sentiments by humanitarian workers resulted in a number of young migrants expressing a lack of confidence in the commitment of these workers to help them deal with the challenges of unemployment and protect them from unfair workplace conditions. Aid workers often did not follow up on the many reported cases of abuse and exploitation: migrant children faced a plethora of challenges at their workplaces, including poor or insufficient remuneration, lack of compensation after completing the work, delays in receiving payment, long working hours, dangerous work (for example, girls employed as domestics were at risk of being sexually abused and exploited), strenuous work, as well as verbal and other physical abuse. Almost every mention of independent migrant girls in Musina during formal situations was prefaced by mentions of the sexual abuse and exploitation they could face at various workplaces. This legitimized calls from service providers to intervene in the lives of migrant girls and protect them by enforcing bans against the employment of people younger than age 18, for example, including those who had finished secondary school and were only interested in doing paid work.

The idea that girls on the move should not work was justified through the argument that it led to delinquent behavior. According to one female aid worker, for example: "Some children doing odd jobs get money but go back to the streets. They buy glue and other drugs. They create another challenge." Anti-work views concerning girls were connected to the general assumption that girls on the move are delinquent and immoral. Attitudes like this, which pathologized and devalued migrant girls' work, resulted in lethargic responses to calls by working children for protection at workplaces or for help in tackling the problem of unemployment among young people. Aid workers' generalizations about the sexuality of migrant girls reinforced prejudices against this population and, consequently, such prejudices led to the marginalization and exclusion of girls perceived to have lost their childhood sexual innocence.

Although some aid workers were aware of the "feminization of migration" (Castles and Miller 1998: 181), the discourse of aid workers in general reinforced patriarchal views of gender, particularly the notion of considering home as "the most appropriate place for women" (Palmary 2010: 53). Zimbabwe was constructed as home and a place of morality. This supports the point made by Bourdillon et al. (2010: 142) that "children become particularly vulnerable when they move illegally across international frontiers." It is noteworthy that aid workers tended to exclude themselves from issues of immorality when they crossed international borders. Aid workers often drew from conservative discourses of "traditional" femininities to cast aspersions on independent migrant girls and further entrench patriarchy by prescribing ways in which "good girls" should behave. Some aid workers gave the impression that the childhood of these migrant girls was unsavory, conceptualizing the migratory movement of girls as a "social rupture" (Hashim and Thorsen 2011: 11) that in the process reproduced the existing gender hierarchy characteristic of the field of migration. For many aid workers, female migration was synonymous with immorality, and this made them at some moments regard the problem of unemployment of young migrants as less than important. This thinking reinforces gender stereotypes perpetuated by other women who were still steeped in the view that a "normal" girl or woman does not migrate or enter "adult worlds" (Stephens 1995: 9) without being accompanied by an adult, because her childhood can be stolen, lost, or corrupted. Thus, what one boy, his peers, and his aid workers saw as an absence of feminine purity resulted in the disqualification of independent migrant girls as suitable candidates for marriage. These findings corroborate those of Boehm (2006: 153) who writes that: "in order to become a complete socially and morally accepted adult being, marriage was, and still is, considered an essential precondition."

Labeling these girls as immoral served to stigmatize and discriminate against them, and justified barring them from accessing workplace protection assistance from aid workers and other service providers. Calling children sexually "loose" opposed the official understanding of children as innocent and reinforced the notion that girls on the move have always been victims of sexual abuse. Associating these girls with immorality and victimization marginalized them in their society and rendered the problem of their unemployment as peripheral. In response to perceived immoral behavior, aid workers drawing from the discourse of child protection crafted interventions in order to control certain behaviors of independent migrant girl children, for example, those pertaining to their sexuality. Aid workers drawing from the discourses of parenthood on the one hand, and patriarchy on the other (seeing women as weak), felt morally obliged to intervene and save these girls' lives. Thus, the discourse in which the girl child needs to be protected was used to strategically "infantalize" migrant girls, including those who were looking for jobs or working in order to leverage some control over their lives. It also belittled the problems associated with the unemployment of these young migrants. The same kinds of argument—concerning immorality, the need for protection, and stigma associated with being on the move—were not emphasized when the discourse turned to migrant boys.

Conclusion

Based on the foregoing ethnographic research, one cannot assume that a uniform discourse regarding children's innocence dominates the representation of migrant children by humanitarian agencies and aid workers. Rather, migrant children battling with the serious problem of unemployment are being represented in multiple ways that have varied consequences in the lives of these children. Aid workers mobilize different and contradictory discourses about these children, depending on the social context. This matters especially because aid workers have shifting interests and understandings of childhood (and especially of girlhood) that are manifest during formal and informal situations in sometimes differing ways. This is not surprising, as childhood is lived and experienced contextually (Prout and James 1990).

The ambiguities and contradictions in the representations of childhood that are manifest within the framework of social and power structures in aid work are reminiscent of Ensor's (2010: 16) observation that "discourses on children and childhood are fluid and evolving." Thus I suggest that perceptions of migrant children's actions as they try to ward off the challenge of unemployment should not be divorced from the social context. In particular, when analyzing the representations of migrant children, it is important to take into account the life-worlds, choices, and actions of different social actors in a difficult social milieu before making conclusions.

As the research presented here shows, paid work is central to independent working children's lives and features prominently in discussions on childhood. There were two dominant discourses: anti-child and pro-child work. The discourse regarding anti-child work, which tended to dominate during formal situations, portrayed migrant children as victims of poverty, forced into work; however, the lived realities of these children sometimes altered the way in which they were represented. Thus, this study points towards the need for ground analysis of the social context in which migrant children work or make the decision to work, as these two competing discourses operate at different times.

During formal situations, aid workers' efforts to protect migrant children tended to be based on the social categorization of children as innocent victims who are not to blame for their situation (see Burman 2008), being weak and too passive to successfully overcome various constraints. This chapter has shown the problems in assuming that independent migrant children looking for jobs or already working are being cast as "bad" or irresponsible by virtue of their behavior, which opposes the dominant discourse of childhood innocence and vulnerability. Depending on the situation, their involvement in these activities could generate positive representations. Discourses of innocence, vulnerability, and children's rights were invoked, often portraying children with much contradiction: as vulnerable yet generally strong, as innocent victims yet perpetrators of social ills including crime, as responsible social beings and yet irresponsible by the nature of their youth, as manipulators and manipulated, and both cultured and uncouth (see Honwana and De Boeck 2005). This resulted in shifting understandings of unemployment and young migrants.

Representations of independent migrant children tend to be moralized and gendered. The social actors negotiate the ideal of a "good" girl child in complex ways, managing the normative expectations of childhood under different circumstances. The pathologization of migrant girls reflects the prevailing international and local discourses of childhood that frame "normal" children as innocent.

The double-speak in the representation of children helped to maintain the status quo at local and global levels, justifying certain practices towards migrant children (interventions or non-interventions in their lives) as well as the pathologization of working migrant girls or the endorsement of work and migration for young females. Importantly, this had the effect of masking a larger set of anxieties about youth unemployment, particularly of females who were often framed as vulnerable and thus better off staying within the protective home environment. Following Check's (2004: 1143) argument, in which the discursive frame that dominates at a particular time is the "consequence of the effect of power relations," this study indicates that different power relations were instantiated by aid workers who had shifting discourses on childhood, girlhood, vulnerability, and child migration. Local and global discourses mobilized by aid workers not only tended to operate at different moments to advance specific interests or "politically correct" ideas regarding each given situation, but also generated different power relations that in turn produced different consequences for the representations of independent migrant girls in the context of unemployment of young foreign migrants.

There was no pervasive climate of intolerance towards child work or employment of young migrants. The line between anti- and pro-work approaches was often blurry and often shifted. The plurality of discourses, often competing, provided a fertile ground for situational and inconsistent representations of working migrant girls and affected the understandings of the unemployment of people in the broad socially constructed category of childhood. Evidence from Musina suggests that the multiple and contrasting discourses of childhood and vulnerability that were circulating among aid workers were producing mixed representations of migrant girls resulting in different consequences for these young migrants as they negotiated the challenges of unemployment and employment.

Acknowledgements

Study results emanate from PhD research funded by the Zeit-Stiftung Ebelin und Gerd Bucerius "Settling into Motion" program, Atlantic Philanthropies, and South

Africa's National Research Foundation (NRF). Funding from the Wellcome Trust under the Migration & Health Project (maHp) and the Wits University School of Governance's Life in the City Project enabled me to write this book chapter.

References

Boehm, Christian. 2006. "Industrial Labor, Marital Strategy and Changing Livelihood Trajectories among Young Women in Lesotho." In *Navigating Youth, Generating Adulthood: Social Becoming in an African Context*. Catrine Christiansen, Mats Utas, and Henrik E. Vigh, eds. Uppsala: Nordiska Afrikainstitutet, pp. 153–182.

Bourdillon, Michael. 2006. "Children and Work: A Review of Current Literature and Debates." *Development and Change*, 37(6): 1201–1226.

Bourdillon, Michael. 2008. "Children and Supporting Adults in Child-Led Organisations: Experiences in Southern Africa." In: *Generations in Africa: Connections and Conflicts*. S. van der Geest, ed. Beyruth: LIT Verlag, pp. 323–347.

Bourdillon, Michael. 2009. "Children as Domestic Employees: Problems and Promises." *Journal of Children and Poverty*, 15(1): 1–8.

Bourdillon, Michael, and Gerd Spittler. 2012. "Introduction." In *African Children at Work: Working and Learning in Growing up for Life*.Gerd Spittler and Michael Bourdillon, eds. Berlin: LIT Verlag, pp. 1–22.

Bourdillon, Michael, Deborah Levison, William Myers, and Ben White. 2010. *Rights and Wrongs of Children's Work*. New Brunswick, NJ: Rutgers University Press.

Boyden, Jo. 2009. "Risk and Capability in the Context of Adversity: Children's Contributions to Household Livelihoods in Ethiopia." *Children, Youth and Environments*, 19(2): 111–137.

Burman, Erica. 2008. "Beyond 'Women vs. Children' or 'Women and Children': Engendering Childhood and Reformulating Motherhood." *International Journal of Children's Rights*, 16: 177–194.

Castles, Stephen, and Mark J. Miller. 1998. *The Age of Migration: International Population Movements in the Modern World*. New York: Guilford Press.

Cheek, Julianne. 2004. "At the Margins? Discourse Analysis and Qualitative Research." *Qualitative Health Research*, 14(8): 1140–1150.

Chirisa, Innocent. 2013. "Social Protection Amid Increasing Instability in Zimbabwe: Scope, Institutions and Policy Options." In *Informal and Formal Social Protection Systems in Sub-Saharan Africa*. Stephen Devereux and Melese Getu, eds. Kampala: Fountain Publishers, pp. 121–154.

Chirisa, Innocent, and Tawanda Muchini. 2011. "Youth Unemployment and Peri-Urbanity in Zimbabwe: A Snapshot of Fessons from Hatcliffe." *International Journal of Good Governance*, 2(2.2): 1–15.

Clarke, John. 2004. "Histories of Childhood." In *Childhood Studies: An Introduction*. Dominic Wyse, ed. Oxford: Blackwell Publishing, pp. 3–12.

Ennew, John, William Myers, and Dominique Pierre Plateau. 2005. "Defining Child Labor as if Human Rights Really Matter." In *Child Labor and Human Rights: Making Children Matter*. Burns Weston, ed. Boulder, CO: Lynne Rienner, pp. 27–54.

Ensor, Marisa. 2010. "Understanding Migrant Children: Conceptualisations, Approaches, and Issues." In *Children and Migration: At the Crossroads of Resiliency and Vulnerability*. Marisa Ensor and Elżbieta Gozdiak, eds. New York: Palgrave, pp. 15–35.

Finnström, Sverker. 2006. "Meaningful Rebels? Young Adult Perceptions on the Lord's Resistance Movement/Army in Uganda." In *Navigating Youth, Generating Adulthood*

Social Becoming in an African Context. Catrine Christiansen, Mats Utas, and Henrik Vigh, eds. Uppsala: Nordiska Afrikainstitutet, pp. 203–227.

Gigengack, Roy. 2008. "Critical Omissions: How Street Children Studies Can Address Self-Destructive Agency." In *Research with Children* (2nd edn). Pia Christensen and Allison James, eds. New York: Routledge, pp. 205–219.

Hashim, Iman, and Dorte Thorsen. 2011. *Child Migration in Africa*. London: Zed Books.

Hassim, Shireen, Tawana Kupe, and Eric Worby. 2009. *Go Home or Die Here: Violence, Xenophobia, and the Reinvention of Difference in South Africa*. Johannesburg: Wits University Press.

Hoffman, Diane. 2011. "Saving Children, Saving Haiti? Child Vulnerability and Narratives of the Nation." *Childhood*, 19(2): 155–168.

Honwana, Alcinda, and Fillip De Boeck. 2005. "Children & Youth in Africa: Agency, Identity & Place." In *Makers and Breakers: Children and Youth in Postcolonial Africa*. Fillip De Boeck and Alcinda Honwana, eds. Oxford: James Currey, pp. 1–18.

James, Allison, Chris Jenks, and Alan Prout. 1998. *Theorising Childhood*. Cambridge: Polity Press.

Jenks, Chris. 1996. *Childhood*. London: Routledge.

Lancy, David. 2008. *The Anthropology of Childhood: Cherubs, Chattel, Changelings*. Cambridge: Cambridge University Press.

Landau, Loren. 2010. "Loving the Alien? Citizenship, Law, and the Future in South Africa's Demonic Society." *African Affairs*, 109(435): 213–230.

Long, Norman. 1992. "From Paradigm Lost to Paradigm Regained? The Case for an Actor-oriented Sociology of Development." *Revista Europea de Estudios Latinoamericanos y del Caribe* [*European Review of Latin American and Caribbean Studies*], 49: 3–24.

Luebker, Malte. 2008. *Employment, Unemployment and Informality in Zimbabwe: Concepts and Data for Coherent Policy-Making*. Issues Paper No. 32 and Integration Working Paper No. 90. Geneva: International Labour Organization. Retrieved from https://unstats.un.org/unsd/gender/Ghana_Jan2009/Background%20doc2%20for%20paper%2039%20(ILO-WP-90).pdf

Mahati, Stanford. 2012a. "Children Learning Life Skills Through Work: Evidence from the Lives of Unaccompanied Migrant Children in a South African Border Town." In *African Children at Work: Working and Learning in Growing Up for Life*. Gerd Spittler and Michael Bourdillon, eds. Berlin: LIT Verlag, pp. 249–278.

Mahati, Stanford. 2012b. "The Representations of Unaccompanied Working Migrant Male Children Negotiating for Livelihoods in a South African Border Town." In *Negotiating Children's and Youth Livelihoods in Africa's Urban Spaces*. Michael Bourdillon and Ali Sangare, eds. Dakar: CODESRIA, pp. 67–86.

Mahati, Stanford. 2015. *The Representations of Childhood and Vulnerability: Independent Child Migrants in Humanitarian Work*. Unpublished PhD thesis. Johannesburg: University of the Witwatersrand.

Mason, Peter R. 2009. "Zimbabwe Experiences the Worst Epidemic of Cholera in Africa." *Journal of Infection in Developing Countries*, 3(2): 148–151.

Meyer, Anneke. 2007. "The Moral Rhetoric of Childhood." *Childhood*, 14(1): 85–104.

Muzvidziwa, Victor. 2001. "Zimbabwe's Cross Border Women Traders: Multiple Identities and Responses to New Challenges." *Journal of Contemporary African Studies*, 19(1): 67–80.

Nieuwenhuys, Olga. 1996. "The Paradox of Child Labour and Anthropology." *Annual Review of Anthropology*, 25: 237–251.

Orgocka, Aida. 2012. "Vulnerable Yet Agentic: Independent Child Migrants and Opportunity Structures." In *Independent Child Migration—Insights into Agency, Vulnerability, and*

Structure: New Directions for Child and Adolescent Development, Vol. 136. Aida Orgocka and Christina Clark-Kazak, eds. San Francisco, CA: Wiley Periodicals, pp. 1–11.

Palmary, Ingrid. 2009. *For Better Implementation of Migrant Children's Rights in South Africa*. Pretoria: UNICEF.

Palmary, Ingrid. 2010. "Sex, Choice and Exploitation: Reflections on Anti-trafficking Discourse." In *Gender and Migration: Feminist Interventions*. Ingrid Palmary, Erica Burman, Khatidja Chantler, and Peace Kaguwa, eds. London: Zed Books, pp. 50–63.

Prout, Alan, and Allison James. 1990. "A New Paradigm for the Sociology of Childhood? Provenance, Promise, and Problems." In *Constructing and Reconstructing Childhood: Contemporary Issues in the Sociological Study of Childhood*. Allison James and Alan Prout, eds. London: The Falmer Press, pp. 7–35.

Raftopoulos, Brian. 2009. "The Crisis in Zimbabwe, 1998–2008." In *Becoming Zimbabwe: A History from the Pre-Colonial Period to 2008*. Brian Raftopoulos and Alois Mlambo, eds. Harare: Weaver Press, pp. 201–250.

Rankin, Neil, and Gareth Roberts. 2011. "Youth Unemployment, Firm Size and Reservation Wages in South Africa." *South African Journal of Economics*, 79(2): 128–145.

Robinson, Kerry. 2008. "In the Name of 'Childhood Innocence': A Discursive Exploration of the Moral Panic Associated with Childhood and Sexuality." *Cultural Studies Review*, 14(2): 113–129.

Rurevo, Rumbidzai, and Michael Bourdillon. 2003. "Girls: The Less Visible Street Children of Zimbabwe." *Children, Youth and Environments*, 13(1): 150–166.

Rutherford, Blair. 2008. "An Unsettled Belonging: Zimbabwean Farm Workers in Limpopo Province, South Africa." *Journal of Contemporary African Studies*, 26(4): 401–415.

Rutherford, Blair. 2011. "The Politics of Boundaries: The Shifting Terrain of Belonging for Zimbabweans in a South African Border Zone." *Africa Diaspora*, 4: 207–229.

Statistics South Africa. 2016. *Quarterly Labour Force Survey. Quarter 3*. Statistical Release P0211. Pretoria: Statistics South Africa.

Stephens, Sharon. 1995. "Child and the Politics of Culture in 'Late Capitalism'." In *Children and the Politics of Culture*. S. Stephens, ed. Princeton, NJ: Princeton University Press, pp. 3–48.

Twum-Danso, Afua Imoh. 2013. "Children's Perceptions of Physical Punishment in Ghana and the Implications for Children's Rights." *Childhood*, 20(4): 472–486.

Valentine, Gill. 2003. "Boundary Crossings: Transitions from Childhood to Adulthood." *Children's Geographies*, 1(1): 37–52.

7 The rise of the J-1 Summer Work Travel program and its rhetorical links to US youth unemployment

Catherine Bowman

On August 17, 2012, the Public Broadcast Service (PBS) program *NewsHour* aired an investigative report entitled "Making Sense of Summer Work Visas for Foreigners." The report was timely since, during the month of its broadcast, the unemployment rate for Americans aged 16–24 stood at roughly 16%, or more than twice the national rate.[1] In the PBS segment the economics correspondent Paul Solman profiled the Summer Work Travel (SWT) program, a category of the US Department of State (DOS) J-1 Exchange Visitor Program. The "J-1" visa is so named for its alphabetical placement in the nonimmigrant, or temporary, visa section of the Immigration Nationality Act (1965). Lawmakers established the J-1 Exchange Visitor Program under the Fulbright-Hays Act of 1961 as a measure to promote public diplomacy during the Cold War. According to the late senator J. William Fulbright (D-AR), a co-author of the act, the program was designed to provide international college students the opportunity to experience American life during the summer while offsetting travel costs with incidental employment (US Senate 1991: 132). In this capacity, SWT program participants have worked in restaurants, hotels, grocery stores, and amusement parks, among other sectors.[2] During the first 30 years of the program, participants enrolled at a modest pace, coming mostly from Western Europe; in 1996 the DOS reported only 20,728 SWT participants.[3] But, by 2008, enrollment in the SWT program had exploded to 152,726 participants annually and enrollment had shifted to Eastern Europe.[4] As illustrated in Table 7.1, in recent years this demographic trend has continued, with growing numbers of participants arriving from Eastern Europe, South America, and Asia (Department of Homeland Security [DHS] 2013; Bowman and Bair 2016).

In this chapter I evaluate why the claim that the SWT program displaces US youth workers has a rhetorical durability even while lacking a conclusive empirical basis. By *rhetorical durability* I refer to the concern expressed by lawmakers (on both sides of the argument), policy analysts, and media that the J-1 SWT program poses a credible threat to young US job seekers.

In 2013 Senator Bernie Sanders (D-VT) pushed to prohibit employment of SWT students and to mandate domestic youth job training programs as part of the Border Security, Economic Opportunity, and Immigration Modernization Act of 2013 (Senate Bill 744), which ultimately did not pass.[5] President Trump has

Table 7.1 The ten largest J-1 SWT sending countries, 2011

Country of Citizenship	Count of SWT Participants
Russia	12,181
Turkey	8,092
Ukraine	7,283
Bulgaria	7,197
Ireland	6,622
China	6,590
Thailand	6,469
Brazil	5,996
Romania	5,522
Poland	3,950

Source: Department of Homeland security (DHS) 2013.

also taken aim at the program. On the campaign trail Trump vowed that, if elected, he would end "employment-based J-1 cultural exchange programs like the SWT visa" (Associated Press 2016). And, shortly after his inauguration on January 23, 2017, a draft copy of an executive order, entitled "Protecting American Jobs and Workers by Strengthening the Integrity of Foreign Worker Visa Programs," called for the country to "reform the J-1 Summer Work Travel program to improve protections of US workers and participating foreign workers" and "prioritize the protection of American workers—our forgotten workers—and the jobs they hold" (Yglesias and Lind 2017).[6] Despite his condemnation of the SWT program, critics like Trump lack a conclusive empirical basis to make such claims, since neither the DOS nor the US Department of Labor (DOL) have performed a formal SWT labor market test.

In this chapter, I make the case that the J-1 SWT program has undergone a significant legal transformation since its inception in 1965; this transformation undermines its purported diplomatic intent and lends certain legitimacy to its critics' charges that it has morphed into a labor scheme exceeding the DOS capacity to effectively monitor its labor market impacts and potential worker abuses. I also call for the DOS to annually publish demographic and labor market data on the employment-based categories of the J-1 visa. Doing so will empower researchers and labor advocates to make evidence-based assessments of the SWT program's impact. Without such transparency, the SWT program remains vulnerable to critique, and US and J-1 workers alike risk potential abuse and displacement.

The relationship between the J-1 SWT program and youth unemployment: data gaps and rhetorical durability

Standing in the poolside sun as he conducted his interview, the PBS correspondent Paul Solman pointedly asked a question of the owner of a swimming pool management company in Manassas, Virginia, who staffed half of his lifeguards using SWT students: "Isn't this [J-1 SWT program] taking jobs away from Americans?" (PBS 2012). Solman's own journalistic investigation turned up a complicated answer. The employers he spoke with could not agree about whether it threatens the job prospects of American young people. One employer, who hired US young people exclusively, cited his competitors' use of international summer workers as the reason he consistently was underbid on pool management contracts and thus unable to hire more US teens. Still other employers invoked familiar tropes about immigrants as the ideal workforce, insisting that US young people are often less reliable, diligent, or willing to do seasonal, low-skilled jobs.[7]

In Solman's discussions with immigrant labor experts and cultural-exchange industry representatives, the response was similarly mixed. Jerry Kammer, a senior policy analyst with the Center for Immigration Studies (CIS), insisted that employers understate the savings to them of hiring SWT participants over US teens. On the other hand, Michael McCarry, a lobbyist for the Alliance for International Educational and Cultural Exchange, contended that the impact on the US youth labor market was negligible considering the size of the program relative to that of the country's economy. Solman concluded his segment on an ambivalent note: "When it comes to the J-1 SWT program [... it generates] goodwill, for sure, but cost[s] at least some American jobs" (PBS 2012).

The fact that Solman failed to reach a definitive conclusion about the SWT program is rather unsurprising given its legal construction as a cultural exchange visa. After all, pursuant to SWT regulations, which were originally written to address cultural exchange, neither employers nor DOS administrators are required to conduct a labor market test before they enter into a contract with international workers.[8] Similarly, the DOS is not required by law to make detailed occupational data on SWT workers publically available. Given this reality, arriving at empirical conclusions about the program is practically impossible; however, various studies shed light on a broader US labor market context in which youth employment rates have been on a steady decline. According to the estimates of labor economist Christopher L. Smith, the employment-population ratio for young people (aged 16–17) fell by one-third in 2011—the lowest youth employment level ever recorded (Smith 2011: 1). As he explains, while it was exacerbated by the 2008 recession it represents a trend that began in the 1990s and deepened into the early 2000s (Smith 2011: 1–2). Some argue that this decline stems from supply-side factors, such as the increased emphasis among young people on college preparation and extracurricular activities over gaining experience in the labor market through summer and part-time low-wage jobs (Aaronson et al. 2006). Other researchers maintain that the increased presence of immigrants has crowded out young workers since adult immigrants and US native teens compete for similar jobs (Camarota and Ziegler 2010).

Smith (2010, 2011) finds that *both* supply- and demand-side factors account for the decline in youth employment. In particular, he identifies statistically significant effects of the increased competition of young job seekers with adult natives displaced by layoffs in the manufacturing sector and by immigrant labor competition (Smith 2011: 4). He also suggests that the long-term effect of immigration on US young people may vary by factors such as race, and that effect may be positive or negative depending on how readily one can translate lost work hours into more schooling, as well as how "sizable and longstanding" the returns are for early work experiences (Smith 2010: 20). Critics of the SWT program suggest that the impact on American young people seeking jobs is especially significant in urban and surrounding labor markets like Boston where SWTs tend to be more heavily concentrated in the city and in surrounding resort areas such as Cape Cod, Martha's Vineyard, and Nantucket (Kammer 2013).[9]

Given what appears to be a connection between the decline in youth employment and the rapid expansion of the SWT program, some lawmakers and policy experts have expressed their concern about its unchecked growth. In August 2011, Senator Mark Udall (D-CO) wrote an open letter to then secretary of state Hillary Clinton, requesting "an outline of the steps that the Department has taken to ensure proper oversight and enforcement to protect against possible misuse of the visa program as it pertains to the protection of *US workers*" (Udall 2011) (emphasis added). In the same year, the Economic Policy Institute—a nonpartisan think tank dedicated to research on low- and middle-class workers—and the CIS both published reports critical of the SWT program. Cataloguing a long list of criticisms, the CIS report specifically noted that it "displaces Americans from the workplace at a time of record levels of youth unemployment" (Kammer 2011: 2).

As previously noted, Senator Sanders successfully proposed a youth jobs amendment to Senate Bill 744. In his address to Congress, Sanders insisted that the SWT program had "morphed into a low-wage jobs program to allow corporations like Hershey's and McDonald's [...] to replace young American workers with cheaper labor from abroad" and thus proposed a measure to counter its effects (Sanders 2013: 4557). Again in 2016, while on the campaign trail, Trump added to the chorus of those decrying the program; he proposed that the "J-1 visa jobs program for foreign youth" be terminated and "replaced with a résumé bank for inner city youth provided to all corporate subscribers to the J-1 visa program."[10]

While the public lacks data on SWT participant demographic characteristics and the industries they occupy (and at what wage levels), the program's detractors rightly hint at the incongruence between the program's original intent and its current unregulated nature. A careful review of its history reveals several pivotal moments that eroded its original intent, leading to a rapid expansion by the middle of the first decade of the 2000s.[11] I argue that these changes have provided fodder for the program's critics. The program's rapid legal transformation through mostly administrative (versus congressional) action has also confounded efforts to adequately regulate it or to make meaningful labor market assessments of its impact.

The history of the J-1 SWT program: from noble beginnings to a dramatic transformation

The history of the SWT program can be characterized by its legal informality as well as its relative obscurity for nearly 30 years—only since the middle of the first decade of the 2000s has the J-1 attracted significant public scrutiny.[12] With the highly politicized nature of many aspects of US immigration policy, the story of the J-1 program generally, and the SWT in particular, reveals a set of nonimmigrant visa categories that were legally crafted not by Congress but instead by two executive federal agencies—the DOS and the United States Information Agency (USIA)—behind the scenes.[13]

The Fulbright-Hays Act (1961) created a broad mandate for educational and cultural exchanges, but it was through federal agency rulemaking that its framework for the resulting programs was outlined.[14] The language of the Act was noble and ambitious, gaining swift approval. Its stated purpose was to "increase mutual understanding between the people of the United States and other nations [...] to strengthen the ties which unite us with other nations" and to "assist in the development of friendly, sympathetic and peaceful relations between the United States and other countries of the world" (US House 1961). Yet the Act made no reference to the SWT program or to the other streams of the J-1 in which low-skilled employment was the primary activity. Instead, it was through regulations authored by the DOS and USIA—the latter a now-defunct federal agency that had responsibility for cultural exchange programs between 1978 and 1999—that the profiles and nature of activities of the various types of J-1 visa holders were defined. The contours of the SWT program were slow to take shape; it would not become a stand-alone legal category of the J-1 visa until 1999 (USIA 1999: 17976).

The early SWT program years

It is nearly impossible to gain a clear picture of the early years of employment-based cultural exchange programs prior to the 1980s, since many of the relevant visa categories—SWT, camp counselor, and trainee programs—were not even mentioned in the Code of Federal Regulations until the early 1980s. Nonetheless, a 1991 letter written by Senator Fulbright to Congress provides some clues about its origins. In it the senator insisted that the Fulbright-Hays Act (1961) was "intended to cover a broad spectrum of educational and cultural activities" including the SWT program (US Senate 1991: 132). He also emphasized that the principle of reciprocity was a cornerstone of the SWT program and provided evidence of its conformity with the Act:

> Beginning in 1972, the Department of State modified the summer work/ travel [SWT] program to require sponsors of the program to create reciprocal opportunities for US students to work and travel abroad in a similar manner. Thus, the reciprocity requirement has been an integral part of the program for the past 19 years.
>
> (US Senate 1991: 132)

Reciprocity would ensure the SWT program functioned as an "exchange" as opposed to a one-way flow of foreign workers into the United States.

Not until 1981 did the USIA's proposed new regulations define for the first time three employment-based categories of the J-1: the trainee, SWT, and camp counselor programs. The USIA explained: "The proposed amendment to the present regulations is expected to provide organizations with effective guidelines and criteria with which to administer practical training use of the J-1 visa" (International Communication Agency [ICA] 1981: 63322). By distinguishing the three programs as "practical" in contrast to the more traditional exchange categories in the popular imagination (such as research scholar, teacher, and high-school student), the USIA tacitly acknowledged the ambiguous relationship between these activities and its public diplomacy objective. The USIA also admitted the divergent and wide-ranging labor activities in which some J-1 recipients would be engaged and the unique set of guidelines necessary to manage these categories effectively. For example, the USIA enumerated several measures to guard against adverse labor market effects. These included the concept of *reciprocity*, which was supposed to ensure that international SWT participants entering the domestic job market were balanced by an outflow of US students to prevent "adverse impact on labor opportunities for United States youth in the 18- to 23-year-old age bracket" (ICA 1981: 63323). They also prescribed a *delayed arrival* for SWT participants without prearranged employment to "give American students who are interested in obtaining summer jobs from two to four weeks [to find such a job] in an uncompetitive environment" (ICA 1981: 63324). Moreover, the USIA regulations called for an even *geographical distribution* of participants to avoid "clustering" in one region that might result in disadvantages for US youth job seekers (ICA 1981: 63325).

Beyond these safeguards, the proposed regulations outlined sponsor requirements regarding particular categories of J-1 visitors. For example, sponsors of J-1 trainees were required to ensure that wages and standard work hours conformed to US legal standards, and to disclose to participants the cost of living and miscellaneous fees they would have to cover (ICA 1981: 63323). Overall, the proposed regulatory language signaled that the SWT, trainee, and camp counselor categories could generate the kinds of immigrant labor abuses to workers that had occurred with past US temporary immigrant labor programs, such as the Bracero Program (Zatz 1993; Massey et al. 2002). The proposed USIA regulations were set to go into force in 1983 (USIA 1983).

The 1990s: rapid growth and controversy over the SWT program

In 1990, the US Government Accountability Office (GAO) published the report "Inappropriate Uses of Educational and Cultural Exchange Visas." Months earlier, Congress asked the GAO to investigate whether J-1 visa holders were performing activities consistent with the Fulbright-Hays Act (1961). The GAO report found many faults with the J-1 visa program. Its authors called into question the SWT, camp counselor, au pair, and trainee categories specifically, concluding that they were "inconsistent with legislative intent" since they were "not clearly for educational and cultural purposes," and as such "dilute[d] the integrity of the J visa and obscure[d] the distinction between the J visa and other visas granted for work purposes" (GAO 1990: 3). The GAO suggested that nonimmigrant labor visas such as the H, L, and M were likely more appropriate for the activities carried out under the employment-based J-1 programs (1990: 9). The report also concluded that the J-1 regulations were "too vague and general" to ensure compliance (GAO 1990: 14). Similarly, the GAO found fault with the management of the J-1 program, citing "major problems and internal control weaknesses" affecting day-to-day operations, including the tracking of sponsor organizations and accurate accounting of the number of J-1 visa holders entering and leaving the country (GAO 1990: 26).

The GAO's lackluster appraisal of the employment-based categories led the USIA to place temporary moratoria on the expansion of the camp counselor and training programs (USIA 1990: 32907). But, less than a year later, the USIA reversed course, claiming that a number of new initiatives underway as a result of the end of the Cold War, including *private-sector* ventures in many former Soviet republics, had "demonstrated the need for expanding exchange activities involving those regions" (USIA 1991: 65991) (emphasis added). While the USIA acknowledged it had yet to resolve the GAO's concerns, it determined that it was in the United States' political and economic interests to lift the moratoria on the expansion of currently designated exchange visitor training and camp counselor programs (USIA 1991: 65991). While this announcement would have no explicit bearing on the SWT program, it laid bare the tension of providing more regulation and oversight of employment-based J-1 visas amidst growing interest in them.

In response to the GAO's findings in 1991, the USIA proposed significant regulatory reforms to the SWT program, but subsequently abandoned them "in light of significant negative comment" (USIA 1993: 15180). The USIA promulgated a new round of regulations in 1993 but the SWT program was conspicuously absent from them. The agency would not formally address the program until 1996, when it issued a Statement of Policy on the subject. In it, the USIA characterized the legal dispute surrounding the program: "The debate [...] occurs entirely along the fault lines that necessarily underlie the intersection of law and policy. The legal considerations of this debate are straightforward, while the policy considerations are less so" (USIA 1996: 13760). The USIA conceded that it did not possess the statutory authority to administer the SWT program, based in particular on the discrepancy between the J-1 statute and the actual activities of participants. Yet it maintained that it "could be made consistent with the intent" by revising program regulations and establishing it as a separate visa category of the J-1 visa, not merely a subsection captured under the amorphous "other educational activities" category of regulations (USIA 1996: 13761).

In the same 1996 policy statement, the agency likewise addressed two original pillars of the SWT program: reciprocity and a preference for students "lacking sufficient funds to enter the United States as tourists" (USIA 1996: 13761). As the USIA explained it, these requirements were initially intended to mitigate the adverse effects on US job seekers and to reserve the SWT experience for international young people of modest means (USIA 1996: 13761). It admitted that these principles had been "seriously eroded with the passage of time" but it offered no remedy for this attrition. In fact, the statement ultimately granted permission to three of four sponsor organizations at that time to "expand both their number of program participants and the countries from which they are selected."[15] The fourth sponsor, which hosted 75% of all SWTs, would be allowed to continue operations at its current levels but would be prevented from expanding its operation based on what the agency characterized as its practice of leaving participants to "their own devices in securing both employment and accommodation" (USIA 1996: 13760). The USIA offered interim yet binding guidelines to sponsors carrying out SWT program management, but these provisions would continue to lack basic worker protections such as local and federal minimum wage requirements.

Then, in 1998 and with little fanfare, Congress granted USIA the legal authority to administer the SWT program that it had previously lacked. The only rationale the Senate Committee on Foreign Relations (CFR) provided for this congressional measure was that the program was "self-financing" and had long made it possible "for students of average means to enter the US on J-1 visas and to work for three months" (US Senate 1997: 38–39). Congress also relaxed SWT recruitment rules by allowing cultural sponsors to enroll participants without the guarantee of a job upon arrival.[16] In 1999, the USIA promulgated regulations establishing the SWT program as its own category in the CFR (USIA 1999: 17976). In so doing, the USIA resolved the "legal considerations" of the SWT debate it had referenced in 1996 by seeking congressional approval to administer employment-based categories it previously lacked. The USIA simultaneously resolved its policy dilemma: by giving the executive agency broad authority to administer the program and not insisting on pre-placement requirements, Congress also signaled that the long-standing principle of reciprocity could be ignored.[17]

These congressional actions, carried out between 1996 and 1999, would set off a massive expansion in the number of both SWT participants and sponsors in the years to come, transforming it from a little-known and modestly utilized program to the most widely used category of the J-1.[18] The SWT program continued to expand rapidly despite concerns about the program raised by the DOS's Office of the Inspector General (OIG) in 2000, a second GAO report in 2005, and a separate 2012 OIG inspection of the DOS bureau responsible for the J-1 SWT and other employment-based categories—all of which noted the questionable oversight and use of J-1 private-sector exchange visas (DOS 2000; GAO 2005; DOS 2012a).

In summary, the convergence of several factors coincided to fuel the SWT program's growth. Most notably was the increased movement of thousands from Eastern and Central Europe who, given new political freedoms after the Cold War, seized opportunities to work and travel abroad (Castles and Miller 2003); and blanket congressional approval for the USIA to expand it and shed former safeguards intended to protect US youth workers. Ironically, by the 1990s, the fundamental diplomatic work of the Act had largely been accomplished;

communism was no longer a clear and present threat. Nonetheless, US government and industry officials deployed the legislation's rhetoric as part of their
strategic marketing abroad as well as political justification for the SWT program
at home. These actions spurred thousands to travel to the US on the visa. As
Figure 7.1 demonstrates, participation rose dramatically in the late 1990s and
early 2000s.

This rapid transformation in the DOS's stance towards the SWT program in
the early 1990s effectively produced a parallel labor market which, as the legal
scholar Janie Chuang asserts, "conceptually and structurally" removes J-1 workers from the protection of US labor laws (Chuang 2013: 277). In so doing, it also
generates unfair labor market competition for US young people, who traditionally filled SWT roles.

As I detail in the next section, the US government's deregulation of the
program in the late 1990s and its decision to maintain its legal construction as a
cultural exchange has given rise to a program that makes the price of hiring SWT
participants artificially low. It also introduces a stream of workers into the US
labor pool that many employers claim offer skills and assets superior to US
youth workers. I will also outline the SWT program's current legal framework
and demographic composition. In so doing, I explain how its current legal

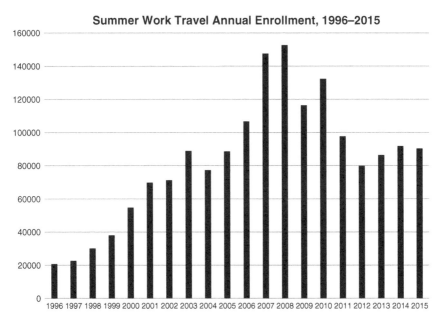

Figure 7.1 SWT program annual enrollment (1996–2015) (1996–2002 enrollment data
from US State Department PowerPoint (on file with author); 2002–2014
enrollment data from Interagency Working Group [IAWG] annual reports;
2015 enrollment data from DOS 2016)

configuration fuels the rhetorical claim that the program is detrimental to American young people.

A contemporary snapshot of the SWT program

As my historical analysis reveals, US regulators have long attempted to negotiate the tension between the SWT program's cultural exchange premise and its practical implications for workers. In fact, a growing body of literature examines the category of cultural exchange worker worldwide and how the label of "cultural exchange" serves to obscure the low-wage labor activities often involved. Social scientists have evaluated how au pairs—another category of cultural exchange workers employed to work in the home as nannies in the United States, Europe, and Australia—are made vulnerable to labor abuse by the labels of "family member" and "cultural exchange participant," respectively (Hess and Puckhaber 2004; Yodanis and Lauer 2005). Similarly, the legal scholar Kit Johnson traces the mobilization of cultural exchange rhetoric by a major US employer, Walt Disney, in order to gain unfettered access to a highly specific pool of migrant labor at a major cost saving (Johnson 2011). While sponsors and the DOS have vigorously defended the cultural merits of the SWT program, its legal construction—unwittingly or not—produces disparate labor market conditions between SWT and other workers.

The SWT program recruitment process

To qualify for an SWT visa, an international applicant must prove he or she is enrolled full-time at an accredited post-secondary educational institution located outside of the United States, has completed one semester of studies, and is proficient in English.[19] SWT participants may reside and work in the United States for up to four months. While a stream of SWT participants arrives nearly every month of the year, the bulk of participants arrive in two waves: June through September and November through February, depending on whether their summer recess aligns with the northern- or southern-hemisphere university schedule.[20]

Eligible international students typically apply to participate through a third-party recruiter in their home country; this agency then partners with a US DOS-designated cultural sponsor organization, which provides international students with various levels of assistance with job and housing placements. Because the DOS does not directly regulate foreign SWT recruiters, their application processes, fees, and practices vary widely.[21] Similarly, there has been no systematic assessment of sponsor organization business practices, industry estimates suggest that sponsors charge an average of US$1,000–4,000.[22] Participants willingly pay these fees because it is the local recruiter—in consultation and coordination with the US sponsor organization—that issues a DS-2019 immigration form; with this document, participants need only successfully pass a US consular appointment to obtain the highly coveted temporary immigration status. Thus, both the discretion of foreign cultural exchange representatives and the incentives for participants are great in these market-based transactions.

This streamlined and mostly privatized immigration process is significant for a few reasons. Because SWT participants are legally treated as purveyors of cultural exchange as opposed to potential temporary migrant workers, the DOL's standards for importing temporary labor—aimed at preventing improper recruitment and protecting workers' rights—do not apply. In the case of the H-2A and H-2B visas—two comparable migrant worker visas used by US employers to staff agricultural and low-skilled nonagricultural positions, respectively—the DOL conducts a labor market test before approving and forwarding them for consular processing. Likewise, under H-2 regulations, it is employers— not migrant workers—who are expected to assume recruitment expenses.[23] While abuses of this provision no doubt occur under the H-2 legal model, under SWT regulations the participant legally absorbs *all* of the recruitment costs.[24] In fact, by its own admission, the DOS does not have the authority to directly regulate SWT employers. On the contrary, the DOS is deeply invested in the privatized cultural sponsor model as a means to amplify its diplomatic reach and fund its own operations.[25]

The legal framework of the SWT program

Relative to other temporary foreign guest workers and US workers, SWT participants enjoy fewer labor protections. Comparative analysis of the SWT with other temporary low-skilled guest-worker programs is relevant not only because it operates in ways that make it functionally analogous to them, but also because it skirts DOL guidelines designed to ensure a true labor shortage. The DOS regulates the J-1 visa categories whereas the DOL regulates all other guest-worker visas and US workplaces generally. Given the DOS's minimal capacity for labor enforcement, it outsources most of its monitoring of SWT employment activities to its designated cultural sponsors. This produces a conflict of interest since cultural sponsors must regulate the same employers they court for business. And, by reporting adverse worker conditions on the job, sponsors place themselves at risk of suspension or termination from the SWT program.[26] In the Hershey SWT scandal of 2011, the cultural sponsor involved—the Council on Educational Travel—allegedly threatened students with termination and deportation when they complained about abusive working and over-priced housing conditions, thereby suggesting that this conflict of interest can result in attempts to silence workers (Breslin et al. 2011). Additionally, unlike H-2 regulations, which require employers to cover certain travel and all visa expenses, SWT employers and sponsors do not pay participants' visa, travel, or consular fees; in addition, participants must purchase health insurance while in the United States.[27]

Similarly, the DOL expressly states that H-2 employer job orders constitute enforceable employment contracts.[28] SWT workers lack such a contract; this places them at a distinct disadvantage should they seek to make a legal claim against their employer (Medige and Bowman 2012; Breslin et al. 2011). Moreover, H-2 workers enjoy a three-fourths contract guarantee.[29] In other words, employers must outline the number of months and hours for which their H-2 worker will be employed. If the employer fails to meet this standard, it must still pay the worker for 75% of the promised time. SWT workers have neither the official contract nor the three-fourths guarantee. Based on anecdotal accounts, many SWT participants in turn suffer both extremes: too much work and too little.[30]

The DOS has taken several steps to address worker abuse. In 2011 it issued regulations that increased pre-employment placement requirements for SWTs prior to their arrival in the United States, called for sponsors to fully vet third parties (foreign recruiting entities and US employers) and SWT job offers, and required sponsors to contact SWT participants on a monthly basis (DOS 2011a: 23178). In November 2011 it issued an SWT cap of 109,000 participants annually (DOS 2011b: 68808). In May 2012, it issued a second set of regulations that prohibited SWT placements in workplaces associated with dangerous working conditions and human trafficking and required sponsors to ensure that participants were exposed to legitimate cultural offerings (DOS 2012b: 25597). In 2015, the GAO issued a report applauding these reforms, but noted the continued lack of transparency around SWT program fees and the ability to verify participant exposure to legitimate cultural offerings (GAO 2015). The DOS issued another proposed rule in January 2017. The proposed rule acknowledged the need to incorporate further safeguards like minimum SWT age, maximum weekly work hours, greater housing protections, and increased sponsor monitoring of employers (DOS 2017). Yet, at the time of writing, organizations like the American Federation of Labor-Congress of Industrial Organizations and the Southern Poverty Law Center (SPLC) have contended that the rule falls short of addressing the program's most problematic elements, such as the sponsor-based enforcement and monitoring structure, high student recruitment fees, the lack of publicly available labor market information on the program, and the continued placement of SWTs in manual labor and hospitality roles—placements that they argue have no discernible cultural value and likely reflect employers' demands to address labor shortages over cultural considerations.[31]

DOS reforms notwithstanding, SWT participants continue to save employers money relative to official guest workers and US workers. Because SWT workers are considered temporary, employers need not pay Social Security, Medicare, or unemployment taxes for them. While immigrant and US workers alike increasingly contend with underemployment, given SWT's lack of contract guarantees and an immigration status contingent on continuous employment, SWT participants can easily become captive to situations of underemployment. This results in SWT participants taking on second or third jobs or returning home in serious debt (SPLC 2014: 13). Similarly, the comparatively low financial risks of hiring SWT participants potentially influence an employer's decision to hire SWT workers instead of US young people.

Conclusion

Given the legal construction of SWT workers as cultural sojourners as opposed to temporary migrant laborers, employers have been quick to add them to their payrolls owing to their cost savings relative to other temporary migrant and native US workers. While geopolitical factors arising from the fall of communism provided a market for importing temporary international youth workers, the SWT program arguably was sustained and expanded based on US employers' growing familiarity with and demand for them. As the historical analysis reveals, the program began as a modest-sized stream of J-1s with an underlying commitment to reciprocity and diplomacy. But, with its reform and expansion in the late 1990s, it presently imports workers who are artificially cheaper and more flexible. Even without conclusive labor market data, it is easy to see why critics raise serious concerns about the growth of cultural exchange immigration and cite US young people as potentially vulnerable to its effects.

A straightforward remedy for the dubious nature of the SWT program flows from the foregoing analysis: the DOS should publish annual systematic data on the employment-based categories. Doing so would bring the DOS in step with the DOL's Office of Foreign Labor Certification legal commitment of publishing similar statistics on other nonimmigrant labor visas, such as the H-2A and H-2B.[32] Such data transparency would allow for a meaningful evaluation of the rhetorical claims about its potentially negative impact on US young people. It could also serve to counter the charge that too much of the program's regulatory framework and oversight lacks transparency and worker safeguards, a charge that would seem to possess merit based on the earlier historical review.

Finally, demographic data on SWT participants made available by the DOS would allow researchers to evaluate the other side of the same coin: that is, the impact of profound labor market restructuring in the post-Cold War era on the employment trajectories of today's international young people, including those who decide to invest in the SWT experience. Too often such phenomena are evaluated in isolation, in this case pitting international and US young people against one another, when their labor market challenges in fact could be potentially interrelated. Such questions may be addressed comprehensively and empirically for the first time if and when the DOS commences routine publication of this increasingly vital J-1 demographic and labor market data.

Notes

1 See Ayres (2013).
2 The list of SWT job venues is based on analysis of a Freedom of Information Act (FOIA) request that details the employers for all of the approximately 90,000 2015 SWT participants. The FOIA request took 19 months to process and was initially denied. To the author's knowledge, this is the first time the State Department has released comprehensive systematic employer information for SWT participants.
3 While 1996 SWT enrollment numbers are not publically available, the author obtained them from a DOS authored presentation on the SWT program (on file with author).

4 The Interagency Working Group 2008 Annual Report provides SWT enrollment numbers. Retrieved from www.iawg.gov/reports/annual (accessed August 14, 2014).

5 Senator Sanders' proposal to prohibit youth employment of J-1 Cultural Exchange participants failed, although his job-training program was adopted. The overarching legislation later failed to come to a vote before the US House of Representatives, causing the bill to expire.

6 The draft order was obtained by Vox, a general-interest news website. Retrieved from https://cdn0.vox-cdn.com/uploads/chorus_asset/file/7872567/Protecting_American_Jobs_and_Workers_by_Strengthening_the_Integrity_of_Foreign_Worker_Visa_Programs.0.pdf (accessed January 8, 2018).

7 Authors such as Hondagneu-Sotelo (2007), MacDonald (2010), and Harrison and Lloyd (2013) provide examples of the desirable immigrant tropes employers invoke when rationalizing their decision to hire based on ethnic, racial, or nationality-based criteria.

8 It should be noted that, in the description of its intended regulatory reforms of the J-1 SWT program published in the *Federal Register* (see DOS 2012b), the State Department did caution SWT sponsors not to place international participants in regions experiencing high levels of unemployment, but outlined no mechanism for complying with this recommendation.

9 For a discussion of the potential regional labor market impacts of the SWT program see Kammer 2013.

10 Trump made comments about the J-1 program in an email to the Associated Press, which was featured in the *Chicago Tribune* article entitled "At Chicago Hotel and Elsewhere, Trump Used Foreign Student Labor He Vows to Ban" (Associated Press 2016).

11 Between May and August 2013, the author carried out a review of hundreds of pages of digitally archived US government documents including: congressional testimony on cultural exchange legislation and related appropriation hearings; Federal Register announcements containing policy statements and updates on administrative rulemaking processes pertaining to the regulatory scheme of SWT and related J-1 programs; annual reports of the DOS and the USIA's J-1 Exchange Visitor Program activities; and internal and independent audits of the J-1 program completed by the GAO and the State Department's OIG.

12 In December 2010, the Associated Press published an investigative piece documenting the harsh living and working conditions of several SWTs; see Mohr et al. (2010). In August 2011, 200 SWT workers walked off the job at a Hershey subcontracting plant in Palmyra, Pennsylvania, complaining of fraud in recruitment, unpaid wages, and retaliation, among other things; see Preston (2011).

13 See Costa (2011) for a detailed description of the transparency concerns inherent to the administrative rule-making process in the case of the J-1 visa categories.

14 For a full discussion of the role of administrative law in setting federal regulations, see O'Connell (2008).

15 In its 1996 announcement, the USIA referenced the American Institute for Foreign Study, YMCA, InterExchange, and Camp Counselors USA as the designated cultural sponsors responsible for managing the four SWT programs it deemed to have adequate screening and monitoring practices in place. It was unclear from the announcement which organization operated the program model that did not guarantee pre-placement in employment or did not assist with housing. Nonetheless, Senator Fulbright's 1991 testimony suggests that the Council for International and Educational Exchange (CIEE) was the fourth sponsor (not referenced) since he characterized it as the biggest and longest-standing SWT sponsor.

16 Pub. L. No. 105–277, an omnibus appropriations bill, and Pub. L. No. 105–244, a Higher Education Act Reauthorization Act, both included the one-line provision: "The

Director of the United States Information Agency is authorized to administer summer travel and work programs without regard to pre-placement requirements." This authority is codified at 22 U.S.C. § 1474.

17 Sec. 405 of Pub. L. No. 105–277.
18 In 2002 the SWT category accounted for 25% of all J-1s, but by 2008 that percentage had increased to 40%; for more discussion of the distribution of J-1 exchange visitor visas, see Costa (2011: 9–11).
19 C.F.R. § 62.32(d).
20 Data obtained by the author from a 2013 FOIA request confirm the concentration of participants during these two cycles.
21 In a December 6, 2013, interview with a representative of the cultural exchange industry, the informant noted that Chinese SWT participants typically paid higher recruitment fees than participants from other countries.
22 Although the fees charged by these organizations vary by agency and country, the author documented this range based on over 30 interviews with present and past SWT participants from Central and South America, Eastern Europe, Asia, and Africa. Variation depended on whether the sponsor arranged employment and travel.
23 20 C.F.R. § 655.135 (j) and (k) pertain to H-2A prohibition on charging employee fees, including recruitment fees and 655.20 (o) and (p) apply to H-2B equivalents.
24 As the Southern Poverty Law Center (SPLC) contends, the discrepancy between law and practice can be great with the H-2A and H-2B; H-2 workers often fall prey to improper and exorbitant recruitment fees. See SPLC (2013).
25 In a 2012 internal audit, the DOS's OIG stated that many staff responsible for administering the SWT and other J-1 private sector exchanges "believe that their jobs ultimately depend on the level of revenues" generated from participant and cultural sponsor fees and that "the Department's reputation can be placed at risk when staff focuses on the quantity of fee-paying participants and sponsors rather than the quality of the visitor's experience." See DOS (2012b: 22).
26 See Costa (2011: 17).
27 22 C.F.R. § 62.14.
28 29 C.F.R. § 501.3(a).
29 20 C.F.R. § 655.122.
30 SWT participants in Myrtle Beach, South Carolina, complained that they worked too few hours to afford basic expenses, which forced some to depart only weeks after arriving in the United States. See Mohr et al. (2010).
31 SPLC and AFL-CIO comments submitted in response to Public Notice 9522. The J-1 Exchange Visitor Program—Summer Work Travel Proposed Rule, Fed. Reg. 82(8) 4120, can be found on the DOS website. Retrieved from www.regulations.gov/dock etBrowser?rpp=50&so=DESC&sb=postedDate&po=0&dct=PS&D=DOS-2016-0034 (accessed January 15, 2018).
32 In July 2014, Representatives Lois Frankel, Jim Himes, and Ted Deutch introduced House Bill 5197, Transparency in Reporting to Protect American Workers and Prevent Human Trafficking Act, with backing from the Economic Policy Institute and the Global Workers Justice Alliance. The bill had yet to be voted on by Congress at the time of this writing.

References

Aaronson, Daniel, Kyung-Hong Park, and Daniel G. Sullivan. 2006. "The Decline in Teen Labor Force Participation." *Economic Perspectives*, Q1: 2–18.
Associated Press. 2016. "At a Chicago Hotel and Elsewhere, Trump Used Foreign Student Labor He Vows to Ban." *Chicago Tribune*, March 14.

Ayres, Sarah. 2013. *The High Cost of Youth Unemployment*. Washington, DC: The Center for American Progress.

Bowman, Catherine, and Jennifer Bair. 2016. "From Cultural Sojourner to Guestworkers? The Historical Transformation and Contemporary Significance of the J-1 Visa Summer Work Travel Program." *Labor History*, 58(1): 1–25.

Breslin, Colleen P., Stephanie Luce, Beth Lyon, and Sarah Paoletti. 2011. "Report of the August 2011 Human Rights Delegation to Hershey, Pennsylvania." Waltham: Brandeis University. Retrieved from www.guestworkeralliance.org/wp-content/uploads/2012/05/Human-Rights-Delegation-Report-Hershey.pdf

Camarota, Steven A., and Karen Ziegler. 2010. "A Drought of Summer Jobs: Immigration and the Long-Term Decline in Employment among US-Born Teenagers." *Backgrounder*. Washington, DC: Center for Immigration Studies.

Castles, Stephan, and Mark J. Miller. 2003. *The Age of Migration: International Population Movements in the Modern World*. London: Palgrave Macmillan.

Chuang, Janie A. 2013. "The US Au Pair Program: Labor Exploitation and the Myth of Cultural Exchange." *Harvard Journal of Law & Gender*, 36(2): 269–344.

Costa, Daniel. 2011. *Guestworker Diplomacy: J Visas Receive Minimal Oversight Despite Significant Implications for the US Labor Market*. Washington, DC: Economic Policy Institute.

Department of Homeland Security (DHS). 2013. "Summer Work Travel Participant Annual Enrollment and Demographic Data for Program Years 2010 and 2011." Unpublished Excel file (obtained under the Freedom of Information Act from DHS; requested February 2015 and obtained July 2016).

Department of State (DOS). 2000. "The Exchange Visitor Program Needs Improved Management and Oversight." Audit Report 00-CI-028. Retrieved from https://oig.state.gov/system/files/8539.pdf

Department of State (DOS). 2011a. "Exchange Visitor Program—Summer Work Travel." *Federal Register*, 78(80): 23177.

Department of State (DOS). 2011b. "Exchange Visitor Program—Cap on Current Participant Levels and Moratorium on New Sponsor Applications for Summer Work Travel Program." *Federal Register*, 76(215): 68808.

Department of State (DOS). 2012a. "Inspection of the Bureau of Educational and Cultural Affairs." Rep. No. ISP-I12-15. Retrieved from http://oig.state.gov/documents/organization/217892.pdf

Department of State (DOS). 2012b. "Exchange Visitor Program—Summer Work Travel." *Federal Register*, 77(92): 27593.

Department of State (DOS). 2016. "Summer Work Travel Participant Employment Placement and Demographic Information for Program Year 2015." Unpublished Excel file (obtained under the Freedom of Information Act from DHS; requested April 2015 and obtained October 2016).

Department of State (DOS). 2017. "Exchange Visitor Program—Summer Work Travel." *Federal Register*, 82(8): 4120–4147.

Harrison, Jill Lindsey, and Sarah E. Lloyd. 2013. "New Jobs, New Workers, and New Inequalities: Explaining Employers' Roles in Occupational Segregation by Nativity and Race." *Social Problems*, 60(3): 281–301.

Hess, Sabine, and Annette Puckhaber. 2004. "'Big Sisters' Are Better Domestic Servants?! Comments on the Booming Au Pair Business." *Feminist Review*, 77: 65–78.

Hondagneu-Sotelo, Pierrette. 2007. *Domestica: Immigrant Workers Cleaning and Caring in the Shadows of Affluence*. Berkeley, CA: University of California Press.

International Communication Agency (ICA). 1981. "Proposed Rule, Exchange-Visitor Program." *Federal Register*, 46(251): 63322–63325.

Johnson, Kit. 2011. "The Wonderful World of Disney Visas." *Florida Law Review*, 63(4): 937–941.

Kammer, Jerry. 2011. *Cheap Labor as Cultural Exchange: The $100 Million Summer Work Travel Industry*. Washington, DC: Center for Immigration Studies.

Kammer, Jerry. 2013. "Figuring the Damage Done by Detached Elites? Add Youth Unemployment to the List." Center for Immigration Studies. Retrieved from http://cis.org/kammer/figuring-damage-done-detached-elites-add-youth-unemployment-list

MacDonald, Cameron Lynne. 2010. *Shadow Mothers: Nannies, Au Pairs, and the Micro-politics of Mothering*. Berkeley, CA: University of California Press.

Massey, Douglas S., Jonge Durand, and Nolan J. Malone. 2002. *Beyond Smoke and Mirrors: Mexican Migration in an Era of Economic Integration*. New York: Russell Sage Foundation.

Medige, Patricia, and Catherine Griebel Bowman. 2012. "US Anti-Trafficking Policy and the J-1 Visa Program: The State Department's Challenge from Within." *Intercultural Human Rights Law Review*, 7: 103–145.

Mohr, Holbrook, Mitch Weiss, and Mike Baker. 2010. *US Fails to Tackle Student Visa Abuses*. Associated Press, December 6.

O'Connell, Anne Joseph. 2008. "Political Cycles of Rulemaking: An Empirical Portrait of the Administrative State." *University of Virginia Law Review*, 94: 889–986.

PBS. 2012. "Making Sense of Summer Work Visas for Foreigners." TV interview by Paul Solman. Retrieved from www.tpt.org/pbs-newshour/video/pbs-newshour-making-sense-of-summer-work-visas-for-foreigners

Preston, Julia. 2011. "Foreign Students in Work Visa Program Stage Walkout at Plant." *New York Times*, August 17.

Sanders, Bernie. 2013. *Border Security, Economic Opportunity, and Immigration Modernization Act*. Congressional Record. Daily Ed. S4557.

Smith, Christopher L. 2010. *The Impact of Low-Skilled Immigration on the Youth Labor Market*. Finance and Economics Discussion Series. Working Paper 2010–03. Washington, DC: Federal Reserve Board.

Smith, Christopher L. 2011. *Polarization, Immigration, Education: What's Behind the Dramatic Decline in Youth Unemployment?*. Finance and Economics Discussion Series. Working Paper 2011–41. Washington, DC: Federal Reserve Board.

Southern Poverty Law Center (SPLC). 2013. *Close to Slavery: Guestworker Programs in the United States*. Montgomery, AL: Southern Poverty Law Center.

Southern Poverty Law Center (SPLC). 2014. *Culture Shock: The Exploitation of J-1 Cultural Exchange Workers*. Montgomery, AL: Southern Poverty Law Center.

Udall, Mark. 2011. "Open Letter to Secretary of State Hillary Clinton Asking for Review of Cultural Exchange Programs." Retrieved from www.markudall.senate.gov/?p=press_release&id=1348

US General Accounting Office (GAO). 1990. *US Information Agency: Inappropriate Uses of Educational and Cultural Exchange Visas (GAO/NSIAD-90-61)*. Washington, DC: GAO.

US General Accounting Office (GAO). 2005. "State Department: Stronger Action Needed to Improve Oversight and Assess Risk of the Summer Work Travel and Trainee Categories of the Exchange Visitor Program (GAO-06-106)." Washington, DC: GAO. Retrieved from www.gao.gov/new.items/d06106.pdf

US General Accounting Office (GAO). 2015. *Summer Work Travel Program: State Department Has Taken Steps to Strengthen Program Requirements but Additional Actions Could Further Enhance Oversight (GAO 15-265)*. Washington, DC: GAO.

US House. 1961. *87th Congress, 1st Session. H.R. 8666 Mutual Educational and Cultural Exchange Act*. Washington, DC: Government Printing Office.

US Information Agency (USIA). 1983. "Final Rule, Exchange-Visitor Program." *Federal Register*, 48(11): 1941–1946.

US Information Agency (USIA). 1990. "Statement of Policy and Notice to Sponsors, Exchange Visitor Training Programs; Policy Statement." *Federal Register*, 55(156): 32907.

US Information Agency (USIA). 1991. "Exchange Visitor Camp Counselor and Training Programs." *Federal Register*, 56(245): 65991.

US Information Agency (USIA). 1993. "Notice of Final Rulemaking." *Federal Register*, 58(52): 15180.

US Information Agency (USIA). 1996. "Statement of Policy. Exchange Visitor Program." *Federal Register*, 61(6): 13760–13762.

US Information Agency (USIA). 1999. "Interim Final Rule. Exchange Visitor Program." *Federal Register*, 64(70): 17976–17977.

US Senate. 1991. *Committee on Foreign Relations, Foreign Relations Authorization Act, Fiscal Years 1992 and 1993*, S. Rep. 102-98, 131–133.

US Senate. 1997. *Committee on Foreign Relations, Foreign Affairs Reform and Restructuring Act 1997*, S. Rep. 105-28, 38–39.

Yglesias, Matthew, and Dara Lind. 2017. "Read Leaked Drafts of Four White House Executive Orders on Muslim Ban, End to DREAMer Program and More." Vox, January 25. Retrieved from www.vox.com/policy-and-politics/2017/1/25/14390106/leaked-drafts-trump-immigrants-executive-order

Yodanis, Carrie, and Sean R. Lauer. 2005. "Foreign Visitor, Exchange Student, or Family Member? A Study of Au Pair Policies in the United States, United Kingdom, and Australia." *International Journal of Sociology and Social Policy*, 25(9): 41–64.

Zatz, Marjorie S. 1993. "Using and Abusing Mexican Farmworkers: The Bracero Program and the INS." *Law and Society Review*, 27(4): 851–863.

8 Irish youth unemployment and emigration, 2009–2014

Eleanor O'Leary and Diane Negra

In 2015, after seven years of post-financial-crash austerity and in light of overall decreases in emigration figures and a more stable economy, government ministers in Ireland swiftly turned to sweeping the losses of the last decade under the carpet and inviting young people to return home, framing their time abroad as advantageous to the economy. Such actions not only overestimated the likelihood of emigrant return but also intentionally ignored the continuing lack of opportunity and precarious forms of employment available to young people in Ireland. Speaking of returning home when emigration levels remained high also underlined the level of denial that successive post-crash governments have exercised in relation to emigration. In a speech in the United States, Enda Kenny, the then Taoiseach (prime minister), called on younger people who had emigrated during the crisis to return home to "be the best at home in their own country" (Kelly 2015). He also spoke of how he wanted young families to return home because their children were part of Ireland's future (although presumably if those children were born abroad their sense of home might be very different to the one the taoiseach was evoking). Rhetorical emphasis on economic recovery and return emigration is part of the consolidated new economic order in which the profit interests of elite transnational corporations serve misleadingly as a gauge of national economic health. This chapter examines how the disjuncture of downward mobility and renewed migration is expressed through selected popular culture forms. Our focus is on young people, who represent the cohort most directly affected by these phenomena in the period 2009–2014.

Background to the crisis

A historically unprecedented period of economic growth in Ireland between 1995 and 2007 gave way, beginning in 2008, to a vertiginous experience of economic contraction. The collapse of the so-called Celtic Tiger boom was identified by the International Monetary Fund as the greatest economic dislocation "of any economy since the Great Depression." Its signal feature was a bank bailout aptly characterized by Nicholas Kiersey (2014: 360–361) as "one of the most spectacularly unjust, and undemocratically decided, transfers of wealth from the taxpayers of an advanced Western nation to foreign bondholders (in this case

German, British, and French banks) in history." The Irish government's decision to guarantee its feckless banking sector was quickly recognized as socially disastrous for a rising generation of young people.

A study by Steven Rattner found that, factoring out tax and social welfare structures, Ireland is the most unequal developed nation (Rattner 2014). Inequality dynamics in the country are impacted by such singular factors as a two-tier health system, which incentivizes private health insurance for those who do not qualify for a state-issued medical card, and the presence (particularly in Ireland's capital) of high-earning employees of low to no tax-paying international technology firms that help to sustain high prices on a range of everyday goods. Such pricing is particularly impactful given that half the nation's households earn between €10,000 to €50,000 per annum while more than 300,000 people earn no income at all (Collins 2013). Irish household debt, in the period we examine, was still close to 200% of disposable income.[1] Severe austerity budgeting was put in place from 2009 and, in 2014, the national debt stood at a staggering €183 billion (Debt Clock 2014).

In the context of these developments, this chapter raises the key question of how discourses about unemployment and emigration function in an environment of neoliberal governmentality that cherishes an "imagining of the free market as exemplary technology of the self" and is devoted to "the essential task of producing a population that is ever mindful of the 'transactional reality' of its existence and thus willing to accept the need for austerity in times of crisis" (Kiersey 2014: 358). While fields ranging from economics to sociology to equality studies have much to contribute in analyzing the recession's social character, media studies offers a unique disciplinary pathway for interpreting recession culture given its focus on the analysis of collective symbolic environments that hold enormous sway in shaping public views. This chapter draws attention to the importance of media forms in providing cues through which key neoliberal norms of self-reliance, adaptation, and enterprise are modelled while criticism of the state is marginalized (Kiersey 2014).

Unemployment and emigration: a dual loss

On the whole, youth unemployment was a more visible and lamented phenomenon in Europe during the recession than in the United States. Indeed, it constituted a pressing subject for post-financial-crash Europe with three summits being convened to address the topic. Its extent varied considerably from nation to nation; 2014 figures reported youth unemployment registered at 57.7% in Greece, 54% in Spain, 22.5% in France, and just 7.7% in Germany (Statistica 2014). A 2013 Oireachtas (Irish government) report noted that there had been a "sharper increase" in youth unemployment in Ireland compared to the rest of Europe; according to the report's statistics it had increased by 19% between 2007 and 2013 (in Europe the increase was 7.5% in the same period) (Oireachtas Library and Research Service 2013). Europe's seemingly greater responsiveness to the problem of youth unemployment may be attributed in part to the scale of the problem, with an estimated

5.5 million jobless young people across Europe, and partly to comparatively more visible protest movements staged in cities whose physical infrastructures, comprising plazas, squares, and physically distinct city centers, lend themselves to marches and rallies in ways the cities of the United States often do not.

Ireland may be a bit of an outlier with respect to the phenomenon of youth unemployment, in part because of the historical relatedness of unemployment and emigration. Previous cycles of acute unemployment problems and emigration imperatives are well remembered, but, even by relatively contemporary standards, the post-2008 cycle was distinctive, as Fintan O'Toole points out:

> In the entire, miserable decade of the 1980s, net emigration was 206,000, a figure seen at the time as a shocking indictment of political and economic failure. In the last five years alone it is 151,000. And most of this emigration is of people between 15 and 44: in 2012 and 2013 alone, we lost 70,000 people in this age group. The percentage of 15- to 29-year-olds in the population has fallen from 23.1 per cent in 2009 to 18 per cent in 2014.
>
> (O'Toole 2014)

Such dramatic levels of emigration are perhaps most comparable to the levels experienced in the 1950s, which is commonly regarded as the darkest decade in 20th-century Ireland owing to extremely high levels of emigration, unemployment, and underemployment. Between 1951 and 1961 approximately 400,000 people emigrated from Ireland and these emigrants were overwhelmingly young; persons under 25 were the most likely to emigrate and girls aged between 14 and 19 years represented the largest percentage cohort of migrants (Delaney 2004). The most recent crisis was at one point well on its way to matching those figures, however, with over 228,000 Irish people leaving after 2009 (the total level of emigration between 2009 and 2014 including those of non-Irish nationality exceeded 405,000) (Central Statistics Office [CSO] 2014/2015). While the overall number of young men emigrating exceeded the number of women in the recession, this did not hold for all categories. In a similar demographic trend to the 1950s, statistics provided by the CSO show that young women aged 15–24 were leaving in consistently higher numbers than their male counterparts (CSO 2015).

Female migration also remained high among university degree holders, who were particularly visible among the unemployed in part due to a cultural aversion to their participation in service economy work. Una Mullally (2014) notes that 29,000 emigrants were university graduates in 2013, up from 20,200 in 2012.[2] CSO figures also show that, while the number of male emigrants with third-level degrees dropped by almost 3% since 2011, the number of female emigrants with third-level degrees remained high in the same period and increased further in 2015 (CSO 2014). This is notable given a larger phenomenon of disproportionate cultural attention paid to male citizens in a recessionary popular discourse marked by the use of phrases such as "mancession" (Negra 2014: 225). This gender bias is most clearly highlighted in Irish films, produced during the recession and discussed later in this chapter, which universally focus on a male protagonist and his experiences of emigration, unemployment, and dislocation.

Compared to the emigration crises of the 20th century, the most notable difference between the post-Celtic Tiger crisis and these past events was the considerable and managed silence on the subject exercised by all government parties (such silence being facilitated by the fact that the two dominant political parties in Ireland—Fianna Fail and Fine Gael—are both conservative). By contrast, the skyrocketing emigration and austerity budgeting of the 1950s inspired heated public debates, multiple government reports, including the Commission on Emigration and Other Population Problems, and four changes in government in the post-war period. The lack of political engagement with the issue of emigration in the period 2009–2014 was for the most part also matched by a lack of public attention. The scale of the crisis made this silence all the more unusual. It is important to note that levels of immigration also remained relatively high during the recession, figures produced by the NERI Institute suggest that net migration in the age category of 15 to 24 (representing the number of those who have gone away and stayed away), in the period 2009–2014, amounted to 81,000 (Healy 2015). Taking into consideration short-, medium-, and long-term migration since 2009, the cultural and social impact of such an enormous movement of people in and out of Ireland received far less attention than it deserved.

Young people, then, were particularly impacted by emigration and austerity in the post-Celtic Tiger era. Ireland's 2015 unemployment rate stood at 9.5% (Ferreira 2015); however, when the numbers of people in receipt of various government payments for attending training courses and back-to-work schemes are also factored in, the figure would have been closer to 20% (Hennigan 2015). The number of people under 26 in receipt of unemployment benefit was double that of the rest of the population (Healy 2015). Bearing in mind the politically embedded character of unemployment definitions and data, we should approach such figures in a skeptical way particularly given the customary role of emigra-tion as a safety valve in Ireland. By leaving in vast numbers, young people reduced the pressure on social services that would otherwise have had to support many more in the ranks of the unemployed. Moreover, their absence also contributed to their political under-representation; their marginalization as a lobby group was dramatically enhanced through emigration. It is no surprise then that the most significant cuts to Jobseeker's Allowance were in payments to those aged 15–24 years. Available research clearly shows that young people are no more likely than any other age group to remain on social welfare and that the main obstacle preventing young people from gaining employment was "the absence of jobs, not motivation" (McCarthy 2014). Yet the government persevered with the rationale that young people needed to be encouraged to relinquish their social welfare payments and "activated" through training schemes.

The cuts to the Jobseeker's Allowance were designed to supposedly force unemployed people under 25 to enroll in training and education courses, although some of those very courses were curtailed or delayed due to lack of funding. The picture darkens further when we take on board developments that tied unemployment benefit eligibility to compulsory work programs. In early 2014, the Minister for Social Protection proposed cutting the benefits of any young people who refused to accept internships allocated by the JobBridge national internship scheme.[3] A few months later, press reports confirmed that "the Government has begun withholding dole payments to jobseekers who refuse to take part in its Gateway job activation scheme" (Rogers 2014). As in the United Kingdom, where those on the dole were sometimes subject to compulsory employment in privately owned businesses including high-profile national brands like Marks & Spencer and Poundland, Irish neoliberal business rhetoric seemed to take an increasingly authoritarian stance towards the "idle" unemployed.

Of course, there is a lot that such unemployment data do not capture—not only emigration but also underemployment and precarious contracts of employment that offer inconsistent, temporary, part-time, and short-term work. Although unemployment figures would slowly begin to decrease after the period we are concerned with here, the lack of quality positions and real opportunities continued to be a considerable challenge. A 2015 European report showed that the number of young people in temporary employment in Ireland had increased by more than 20% between 2004 and 2012 (National Youth Council of Ireland 2015). Unpaid internships, zero-hours contracts, and low pay were highlighted by the National Youth Council of Ireland (2015) as comprising a major concern in the report, which argued that "young people are on the front line in coping with the rise of precarious employment."

Unemployment and emigration statistics also fail to represent other attendant phenomena including a national mental health crisis in which youth suicide factored strongly. (It remains customary in the Irish press not to make known a cause of death in cases of suicide.) One particularly painful element of the lack of opportunity for recession-era Irish youth (sometimes referred to as Celtic Tiger Cubs) was the puncturing of the fantasy that they had transcended such historical problems. Those factors that mask unemployment are often culturally specific and in Ireland the most important element in this regard is emigration. The Irish national context is one in which the right to transition to adulthood in your own country is not historically justified, and the recession saw the revival of folk practices such as "the American wake," which was once assumed to have been left behind and associated with permanent emigration.

Privileging masculinity in the popular culture of emigration

Recessionary Ireland was marked by a popular culture that had to make some acknowledgement of youth unemployment/emigration if it was to be socially credible, and yet it has also tended to do so in particular ways so as to sustain the paradox that Ireland is "open for business." This credo organizes public life and the actions of government on numerous levels. The only thing that successive Fianna Fail and Fine Gael governments since 2007 have shown themselves to be unswervingly dedicated to is the low corporate tax rate of

12.5%. (In reality many international corporations in Ireland pay much less than this rate, or indeed nothing at all.) In addition to the importance of not acknowledging or disclosing information that might contest the image of a "business-friendly" Ireland, popular culture tended to revert to very rigid gender norms of enterprising, adjusting men and abiding women. While an emphasis on "crisis masculinity" emerged in Ireland (similar to that in the United States where recessionary concerns quickly came to center upon the social impact of job loss on men), such discourses have been directed towards midlife men rather than young men, part of a larger process in which job loss is culturally understood as a loss of *male* agency. Emigration was not entirely evaded and it really could not be given its scale and scope. The subject was taken up in forms of national media including documentaries, films, and talk shows, while *The Irish Times* newspaper put in place a longstanding series entitled "Generation Emigration."

Emigration and underemployment strongly color a film like *Standby* (2014) in which 28-year-old Alan is caught in limbo, living with his father and working below his skill level at an airport tourist information desk beside his mother, who we assume arranged the job for him (Burke and Burke 2014). Emigration, unemployment, and underemployment are consistently referenced in the film as Irish Alan and American Alice share a night in Dublin, reconnecting eight years after they first met and fell for one another in the United States, when Alice proposed a green-card marriage and Alan declined. The film implies that Alan, who we learn was fired from a banking job, made a mistake in returning to Ireland; when the couple recommit to each other in the conclusion Alan says (in the last spoken words of the film): "I don't have to stay here" (Burke and Burke 2014). At another point, the two encounter several engineering students at a house party who glumly prognosticate about their post-graduation future: "No jobs, probably have to emigrate" (Burke and Burke 2014). In the film's vision of Ireland as a nation with no opportunity for the young we see that even the rituals of youth are now being performed by the elderly, as in their travels through the city Alan and Alice wind up attending the wedding of two septuagenarians. As we will see, *Standby*'s romantic emigration contrasts forcibly with other films we discuss here.

More characteristic of mainstream post-crash media was a studious avoidance of the altered realities of Irish life altogether or the recuperative ideological maneuvers of a "bromance" like *The Stag* (2013), in which a group of male friends rediscover an essential Irish masculinity in a film plot much indebted to the US *Hangover* franchise (Butler 2013). Notably at the film's conclusion a character known only as "The Machine" celebrates a re-secured culture of male affiliation by singing U2's "One" and dedicating the song to his stag party friends with the words "I love those hombres." Taking up the microphone he gives a speech that at once acknowledges and depoliticizes the duress of recent times and sets the stage for recovery, characterizing "One" as:

> also a story about Ireland. And the *men*, and the women of Ireland. In recent times we've taken a hell of a beating. What with the economy, and Europe

tearing us a new one and the Church being total assholes about everything. But we've gotta forgive ourselves, forgive each other and learn to love ourselves again because the thing is we're Ireland. And that, my friends, is deadly.

(Butler 2013)

Emigration has long been a representational concern in Irish popular culture with films like *Korea* (1995) mapping the terror and dread of enforced departure, and with others such as *Gold in the Streets* (1997), *2 x 4* (1998), and *I Could Read the Sky* (1999) evaluating the post-departure experience. Not surprisingly, in the 2009–2014 period it became a subject of increased representational focus in films like *The Omega Male* (2014), *Out of Here* (2013), and *Get Up and Go* (2015). The protagonist of *Out of Here*, Ciaran, a 20-something man who has spent a year traveling outside of Ireland, returns to a betwixt and between existence, finding that his former classmates and family have ambiguously moved on, although his father has become unemployed (Foreman 2013). Ciaran discovers that the ease of travel cannot compensate for the disjuncture of leaving. Far from the stereotypical pomp of the emigrant's return, Ciaran experiences a low-key reintegration, traveling from the airport as the camera picks out banal details of the urban landscape and then arriving at his family home where his mother expresses surprise to see him because she expected him on another day. Ciaran has no exciting stories to tell, and when he does finally unburden himself to a group of peers near the end of the film, he relates forlornly how he met a woman while traveling, but when the two spent the night together he woke only to find that she had robbed him of his belongings.

Such films appeared as distinct manifestations of broader efforts to manage the cultural psychology of downward mobility while they also staged, in both comic and dramatic modes, experiences of unemployment and emigration that markedly impacted post-Celtic Tiger social life. Another prevalent form of expression for the emotional culture of the era was the "surprise homecoming video" that flourished on YouTube in such well-watched forms as "Irish Mums [*sic*] Reaction to Surprise Visit from Her Son," with nearly half a million hits by 2015, and "Mother Is Reunited with Her Daughter after Three Years." (With a nearly comparable number of views, the latter captures an incident on the national television institution *The Late Late Show* in which host Ryan Tubridy first quizzed a mother in the studio audience about what she missed about her daughter, then reunited them on air.[4]) The popularity of such clips can be gauged according to the fact that, relative to the Irish population, they draw a notably high number of views, with the number of hits in each case amounting to roughly 10% of the national population. These surprise homecoming videos also can be read as works of emotion in which the act of leaving is nullified by the ecstasy of return. The cathartic moment of reunification of the family (significantly associated in many of the videos above with the traditional family holiday of Christmas) is often critiqued in the comment sections in relation to authenticity (O'Leary and Negra 2016). The videos typically privilege the emotional reaction of the mother—her performance signals the significance of the event for the family while the children are often amused by her "overreaction" to the return of the grown-up child. In contrast, the representation of return in *Out of Here* underlines both the parents' and Ciaran's apparent loss as to what to do next. When Ciaran asks to borrow money until he can find some work his mother replies with a puff of amusement: "Good luck with that" (Foreman 2013). The videos of return, on the other hand, avoid the social realities of emigration, and there are unsurprisingly few videos of Irish mammies saying goodbye to their children after their holiday visit.

Almost all of the representations of emigration and return discussed above center on a male protagonist and their experience of emigration. In this regard they might be seen to perpetuate a long tradition of male-centeredness in the Irish cinema. As Debbie Ging (2013: 10) has observed, indigenous film in Ireland "is heavily male-dominated, the vast majority of films made—with many notable exceptions—tend to have either been about men or have been told from a male perspective." (Exceptions to the male-centeredness of emigration representation concentrate on the figure of the Irish mammy whose emotional reaction informs the tone of the videos of return described above.) It is worth noting, however, that the most watched viral videos of return almost always focus on the return of a son (rather than a daughter) to a surprised mother figure. In *Out of Here*, Ciaran and his unemployed father circle aimlessly in a world occupied by women who are surviving, abiding, and thriving. Ciaran and, to a lesser extent, his father are encapsulated in long contemplative silences throughout the film; isolated in quiet spaces away from the chaos of the city, work, relationships, and reality. By comparison, Ciaran's mother, sister, ex-girlfriend, and secondary love interest are busy going places and making plans; they have jobs and are surrounded by people and action. This isolation of the male in the recessionary landscape is quite literally represented in Jon Hozier-Byrne's 2014 short film *The Omega Male*. As the last man in Dublin, Conor films himself, walking around the empty city as he describes the freedoms and sense of liberation he enjoys as its final occupant. His realization that everyone else "had just left" comes when he is the only person to turn up at the social welfare office to collect his benefits. The closing scenes act as a counterpoint to the optimism the character expresses throughout the film regarding his situation, when he admits that his short films are difficult to make because he has to do everything himself. His final line to the camera—"Cut, that's a wrap on Conor"—poignantly underlines the nature of his isolation as he rises from a bare mattress on the floor to turn off the camera. Although cultural texts such as *Out of Here* and *The Omega Male* directly address the issue of emigration in a range of contexts they do so within a closed set of highly gendered codes, suggesting that the experience disproportionally disadvantages male citizens. This over-representation does not correlate with the statistical realities of post-crash emigration in Ireland, which showed only a small margin of difference between the levels of male and female emigration and, as highlighted earlier, in the age bracket of 15 to 24 years, female emigration outstripped male emigration every year in the period 2011–2014.

The Gathering: managing emigration, marketing return

In contrast to the depictions discussed above, many representations of emigration were oblique as well as politically and economically decontextualized. In what public relations personnel, tourism industry chiefs, and the government subsequently celebrated as a masterstroke, in 2013 the Irish government sought to incentivize tourism under the auspices of a marketing campaign called "The Gathering," which celebrated bringing family members and others affiliating with Irishness "back home" to Ireland. It was a mark of the recessionary turn that this plan emerged from a government that only 10 years prior (flush with the spirit of Celtic Tigerism) sought to police the boundaries of Irishness through legislation that narrowed previously expansive definitions of national citizenship just as Eastern European and African immigration was surging. Thus dynamics of labor, tourism, and travel in Ireland were increasingly marked by a paradox in which strenuous efforts were made to entice visitors in, even as the nation returned to sending its own citizens out. Given that youth emigration in the period under analysis tended strongly towards long-haul destinations such as Australia, Canada, and China, a fantasy of easy, affordable return also underpinned The Gathering in crucial ways. Event tourism of this kind functioned as simultaneous acknowledgement and denial of the return of mass emigration to Ireland.

One crucial rhetorical strategy adopted by The Gathering was that historical emigrations from Ireland should be valued as diasporas while the character of recent and ongoing emigration was left unspecified. The promotional video *The Gathering—Be Part of It* exhorts the Irish "at home" to invite a vaguely defined group of relations (near and far) and seemingly anyone affiliated with Irishness to make a visit to Ireland. National crisis was papered over in favor of a global reunification narrative including the following evocation: "The Irish Diaspora. A global family of over 70 million people. A people who though scattered far from Ireland, never have Ireland far from their hearts" (The Gathering 2014b). The compulsory nature of such boosterism was starkly revealed when the Irish actor Gabriel Byrne (who had been serving as a global cultural ambassador for Ireland) criticized The Gathering as a scam and was excoriated for it by Leo Varadkar, then Minister for Sport, Tourism and Transport and a rising Fine Gael star (who subsequently moved on to the more prestigious Ministry of Health and is now taoiseach). Byrne, Varadkar suggested, was an aging matinee idol whose views could be safely dismissed since he was of interest only to "women of a certain age" (*Irish Independent* 2012). The minister's ageist and sexist dismissal of a celebrity figure, more often touted as an exemplar of the nation's global artistic achievements, illustrated that these would be trumped by the need to deal punitively with ideological breaches of code at a difficult moment in national economic life.

Waxing lyrical on the rewards of return, The Gathering sought to fully divorce the current wave of emigration from its political and social implications and generally to hold the phenomenon at arm's length. As with the viral surprise-homecoming videos highlighted above, the moment of return was managed and marketed as an occasion when the experiences of loss would somehow be redressed. The media texts designed to accompany The Gathering cautiously avoided the word *emigration* and instead attempted to frame the process of return as a series of events relating to friendship, kinship, long-lost relations, sharing stories and connecting lives, and culture within a global Irish community. Narratives of return that relate to historical emigration are easier to negotiate and decontextualize. Gaelic-speaking Irish Americans, redhead conventions, lace-making, long-lost cousins, and the Tralee Roses of the famed beauty pageant are all woven into stories of connection and celebration of an essential Irishness that cannot be repressed by space and distance. Such essentialism is clearly represented in *The Gathering Magazine*, which reflected on some of the events of the yearlong campaign (The Gathering 2014b: 8). A section of the publication outlined the parameters for "What it means to be Irish," noting that:

> Being Irish is claiming not to watch the Eurovision but secretly knowing every one of our winning entries. It means quoting Father Ted, calling crisps Tatyos, knowing "no" means "yes" to a cup of tea and being part of the infectious laughter that spews from our pubs.
>
> (The Gathering 2014b: 7)

So essential are the representations of Irishness in this section that it threatens to isolate the very generations of emigrants that it claims to welcome back with open arms. Two pages of the magazine focus on Gaelic Athletic Association (GAA) teams, one hurling team based in Bunclody and a camogie team based in Toronto. Here the monumental effort to gloss over the recent escalations in emigration and unemployment almost entirely breaks down. The "Clash of the Ash" page documents the return of almost 50 hurling players and 100 supporters from Australia to take part in a testimonial match against their old home team (The Gathering 2014b: 8). Dislocation and loss are yet again framed within discourses of return and renewal, and the event is warmly described in the following terms: "every crevice of the tiny town seemed alive [...] as the sons and daughters of Bunclody returned to the area from Australia" (The Gathering 2014b). These are sons and daughters in the very literal sense, which is even more clearly represented in the accompanying video in which young people talk about the excitement of being back and other young and older people talk about the joy of having the players come home. The devastation of these losses is perhaps more viscerally rendered in an image from the *Connacht Tribune* in August 2014 (see Figure 8.1), which displays the Carna-Caiseal team from County Galway with eight players from the 2005 team blacked out (Tierney 2014) in an unconventionally brutal representation of the impact of emigration and migration on the team and community.[5]

Another section in *The Gathering Magazine*, entitled "Toronto Ladies Gather in Croker," shared the story of the recently formed Toronto Camogie Club, which went on to win the North American Championships (The Gathering 2014b: 20). In the video piece that accompanies the article, one player details the setting up of the

Figure 8.1 Absent members of the 2005 Carna-Caiseal team (Tierney 2014)

camogie club, stating that "there was a Toronto hurling club in the city and there was no camogie club, with the huge emigration influx of the Irish community we decided that maybe some of the girls wanted to play the sport" (The Gathering 2014a). The story of the club, founded by recently arrived migrants, underlines varying levels of gender disenfranchisement and nostalgic attachment to Irishness in the repeated admission that most of the women had never played hurling in Ireland because of a lack of facilities in their local town or a lack of familial interest in women's sports. The closing paragraph of the accompanying article focuses on a quote from one of the players who says: "You don't realize the emotion of coming home till you've lived abroad. It may sound silly, but this has meant so much to us" (The Gathering 2014a). By framing the moment of return as an event that the entire nation could "be part of," The Gathering attempted to commercially exploit the returned migrant in the precise moment when levels of post-Celtic Tiger emigration peaked. Its seeming success further supports two of the central discussion points in this chapter—the first being the political and public silence regarding the overwhelming levels of emigration from Ireland after the global financial collapse, and the second being the presentation of the ease and ecstasy of return as a dominant discourse on emigration in the post-Celtic Tiger era (O'Leary and Negra 2016).

We're not leaving: popular resistance

A noteworthy activist response to this state of affairs came from the group called "We're Not Leaving," which campaigned against "forced emigration," precarious employment, precipitous increases in university fees, and associated problems of alcoholism, housing, and mental health with the following credo: "Young people will be targeted and hit while we're a soft target. It's time to stand up for ourselves—to get organised and be effective." In an effort to combat and contest euphemistic national rhetoric, We're Not Leaving's Galway branch produced a sharp rejoinder to the 2014 St. Patrick's Day-timed recovery narrative/promotional video *Inspire Ireland*, sponsored by Failte Ireland (The National Tourism Development Authority). Using similar imagery and information, the We're Not Leaving video responded to the boosterism of the *Inspire Ireland* text and critiques it as over-hopeful, diasporic, and corporate-centered. Its closing tagline states: "We're not buying Brand Ireland, We're not going anywhere, We're Not Leaving" (We're Not Leaving Galway 2014). In more individualized ways, others such as Manus Lenihan also responded in provocation to the Failte Ireland piece by producing their own assemblages of images and facts to counter the original video's rosy view. Lenihan's *A St. Patrick's Day Message About the Real Ireland* celebrates a history of engagement and activism and notes:

> Here in Ireland there's a lot to be proud of, and I mean besides our "flexible workforce" [at this point viewers witness an image of a man with a bullhorn beside a sign reading "scrap zero-hour contracts"] and all that guff about "recovery."
>
> (Lenihan 2014)

A photograph posted on We're Not Leaving's Facebook page (see Figure 8.2), which shows an advertisement for a demonstration in October 2013 in Dublin, similarly encourages young people—for whom government cuts to the dole leave "NO CHOICE" for the country's young but to leave—to show up and make their voices heard.

As we have documented here, new rhetorics of emigration in Ireland emerged in the years following the global financial collapse and dramatic "death" of the Celtic Tiger in 2007. Ireland, of course, has maintained a high though also cyclical rate of emigration for centuries, and Irish national identity is diasporic in many respects. Emigration has long been correlated with opportunity, and for Irish middle-class families it may be far more socially acceptable to say that a child has gone to Australia or Canada than to say they are working in a grocery store or a Starbucks. In some respects this wave of unemployment-driven emigration resembles earlier cycles but the key distinguishing feature of the recent phase is that it played out in an era of globalization facilitated in crucial ways by technology. Mobility is a fraught concern in this era as new cultural/ financial rhetorics hype the mobility of capital and the movement of elite actors while enforcing unwanted mobility on others. As Thomas Birtchnell and Javier Caletrío write:

Figure 8.2 On the *Dáil* queue (We're Not Leaving 2013)

Mobility is inextricably tied up with power, inequality, stratification, govern-ance and decision-making—it is about the "capacity to move" being engen-dered by a top-down social movement encouraging an ongoing transition to heightened movement whatever the social cost, including displacement, itinerancy and dispossession.

(Birtchnell and Caletrío 2014: 6)

Thus, with emigration becoming an increasingly more pronounced feature of the Irish social landscape in the period 2009–2014, there was an emergent split between mainstream and state media coverage of the phenomenon and the treatment by alternative and activist media. While coverage by national broadcaster RTE in particular tended to be neutral or approbatory, a range of other Irish print, video, and film material chronicled the social devastation that unemployment and emigration wrought in dramatic, comedic, and satirical modes. The satirical *Waterford Whispers News*, for instance, reported in June 2014 that the Minister for Education had announced a new subject in the Irish Leaving Certificate (the examination that culminates Irish secondary education). The subject of Emigration, the article claimed, would now become a core focus in the curriculum with the study of Australian culture and geography replacing study of the Irish language to better prepare graduates for their likely futures (*Waterford Whispers News* 2014a). In August 2014, on the day that all Irish second-level students received their final results, *Waterford Whispers News* published a follow-up article entitled "The Top Five Destinations You Just HAVE to Emigrate to Post Leaving Cert." Claiming sponsorship by the Department of Social Welfare, the article suggested a range of destinations, depending on the level of points gained in the state examination, listing Southeast Asia, South America, Australia, Canada, and the United Kingdom (*Waterford Whispers News* 2014b). In respect of the latter the article suggested: "You know when you go into a fast food restaurant in Ireland and realise everyone is foreign? Yeah, well basically that's what you'll be doing over there" (*Waterford Whispers News* 2014b).

Conclusion

In a national framework marked by draconian austerity and tremendous public forbearance regarding a bank bailout that even the former US treasury secretary Timothy Geithner characterized as "unaffordable" and "stupid," it is critical to recognize the mechanisms through which the state disassociated itself from and disavowed the new precarity.[6] Youth unemployment and emigration, although historically well rehearsed in Ireland, nevertheless need to be ideologically justified, particularly for a new generation that had expected to transcend such problems. In the period analyzed here, popular culture material, ranging from journalism and commentary to film and viral videos, played a decisive part in establishing and contesting social norms around the downward mobility and despair of young Irish citizens. Part of that process related to a resolute suppression of emigration as a female experience. Situating the new emigration as continuous with a long line of challenges to Irish masculinity under formation, popular culture texts (and particularly film) participated in a process of naturalizing the social and economic devastation of recent years as historically inevitable for a rising generation of Irish men.

The relatively low levels of public protest in Ireland, the failure to prosecute white-collar financial crime, and the extraordinary devotion to a "business-friendly" ethos must be understood in relation to the fact that, as Fintan O'Toole (2014) points out: "The thinning numbers of the young have been unable to mount any sustained challenge to the self-serving orthodoxies of their elders." Cultural protocols stipulating that emigration was painful but necessary, that it would be of short duration and a beneficial experience of youthful self-discovery, or that it could be easily managed via new communication technologies that "erase distance" were rooted in popular discourse. They are part of a horizon of "diminishing expectations" (to repurpose a phrase from the economist Paul Krugman),[7] and they require both identification and analysis to better decipher a popular mindset too often misunderstood as oblivious or passive.

Notes

1 See: *New York Times* 2014.
2 Mullally comments: "That's a whopping increase in just-graduated students leaving within a tiny 12-month period. Obviously, new graduates are emigration-ready. It's easier to leave when most of your possessions fit into a rucksack and you probably don't have the bank screaming at you for mortgage payments and are less likely to have kids screaming at you for everything else. But it's deeply troubling."
3 The JobBridge scheme was widely held in low regard, seen by many as a project designed to camouflage true national unemployment figures and furnish cheap labor to the companies with whom JobBridge contracted.
4 This gambit was a popular one on *The Late Late Show* with four families being reunited on air with loved ones who had gone abroad.
5 These players contested the 2005 county final and are now living in the United Kingdom, the United States, or Australia. Two more have moved to Dublin and are unavailable to play for the club.
6 Geithner's remarks were made public as advance excerpts of his book *Stress Test: Reflections on Financial Crises*. See Kelpie 2014.
7 See Krugman 1997: xi. The central concern of the book is with "an era in which our economy has not delivered very much but in which there is little political demand to do better." Krugman has been a regular critic of Irish austerity measures.

References

Birtchnell, Thomas and Javier Caletrío. 2014. "Introduction: The Movement of the Few." In *Elite Mobilities*. T. Birtchnell and J. Caletrío, eds. London: Taylor & Francis, pp. 1–2.

Burke, Rob, and Ronan Burke (directors). 2014. *Standby* [Motion Picture]. Ireland: Blacksheep Productions/Juliette Films/Paul Thiltges Distributions.

Butler, John (director). 2013. *The Stag* [Motion Picture]. Ireland: Irish Film Board/Treasure Entertainment.

Central Statistics Office (CSO). 2014. "Estimated Migration Classified by Sex and Education Attainment, 2009–2014." Retrieved from www.cso.ie/en/releasesandpublications/er/pme/populationandmigrationestimatesapril2014/#.VHyL59KsVhQ

Central Statistics Office (CSO). 2014/2015. "Table 2 Estimated Migration Classified by Sex and Nationality, 2009–2014, 2010–2015." Retrieved from www.cso.ie/en/releasesandpublications/er/pme/populationandmigrationestimatesapril2014/#.VHyL59KsVhQ

Central Statistics Office (CSO). 2015. "Table 4: Estimated Migration Classified by Sex and Age Group, 2010–2015." Retrieved from www.cso.ie/en/releasesandpublications/er/pme/populationandmigrationestimatesapril2015

Collins, Michael. 2013. "What Is a 'High Earner' in Ireland Today?" *The Journal.ie*, January 17.

Debt Clock. 2014. Finance Dublin, December. Retrieved from www.financedublin.com/debtclock

Delaney, Erica. 2004. "The Vanishing Irish? The Exodus from Ireland in the 1950s." In *Ireland in the 1950s: The Lost Decade*. D. Keogh, F. O'Shea, and C. Quinlan, eds. Cork, Ireland: Mercier Press, pp. 77–89

Ferreira, Joana. 2015. "Ireland Unemployment Rate." *Trading Economics*, September 29. Retrieved from www.tradingeconomics.com/ireland/unemployment-rate

Foreman, Donal (director). 2013. *Out of Here* [Motion Picture]. Ireland: Stalker Films.

The Gathering. 2014a. "The Gathering, Ireland." Retrieved from http://viewer.zmags.com/publication/7afcf26c#/7afcf26c/8

The Gathering. 2014b. *The Gathering Magazine*. Retrieved from http://viewer.zmags.com/publication/7afcf26c#/7afcf26c/8 p. 8 (accessed December 4, 2014).

Ging, Debbie. 2013. *Men and Masculinities in Irish Cinema*. Basingstoke: Palgrave Macmillan.

Healy, Tom. 2015. *Emigration Has Taken Its Toll*. Nevin Economic Research Institute. Retrieved from www.nerinstitute.net/blog/2015/07/03/emigration-has-taken-its-toll

Hennigan, Michael. 2015. "Ireland: Official Unemployment Rate at 9.8% in May; Broad Rate at 19%—444,000 people." Finfacts Ireland, Business and Finance Portal, June 4. Retrieved from www.finfacts.ie/irishfinancenews/article_1028870.shtml

Hozier-Byrne, Jon. 2014. *The Omega Male* [Short Film]. Ireland: Stoneface Films.

Irish Independent. 2012. "Why Men of a Certain Age Must Really Mind Their Manners, Leo." *Irish Independent*, November 11. Retrieved from www.independent.ie/lifestyle/why-men-of-a-certain-age-must-really-mind-their-manners-leo-28895120.html

Kelly, Fiach. 2015. "Kenny Calls on Young Irish in US to Go Home." *Irish Times*, September 25.

Kelpie, Colm. 2014. "Ireland Stupid to Guarantee Banks'—Timothy Geithner." *Irish Independent*, November 12. Retrieved from www.independent.ie/business/irish/ireland-stupid-to-guarantee-banks-timothy-geithner-30738356.html

Kiersey, Nicholas J. 2014. "Retail Therapy in the Dragon's Den: Neoliberalism and Affective Labor in the Popular Culture of Ireland's Financial Crisis." *Global Society*, 28 (3): 356–374.

Krugman, Paul. 1997. *The Age of Diminished Expectations: US Economic Policy in the 1990s* (3rd edn). Cambridge, MA: MIT Press.

Lenihan, Manus. 2014. "A St. Patrick's Day Message About the Real Ireland" [Online Video]. Retrieved from www.youtube.com/watch?v=IXLVClMOjcM.

McCarthy, Daragh. 2014. *Examining the Need to Reduce Jobseeker's Allowance for Young People*. NERI Research. Retrieved from www.nerinstitute.net/blog/2014/09/17/youth-unemployment-and-changes-to-jobseekers-allow

Mullally, Una. 2014. "Workplace Has Become a Terrain of Insecurity and Exhaustion." *Irish Times*, September 1. Retrieved from www.irishtimes.com/news/social-affairs/workplace-has-become-terrain-of-insecurity-and-exhaustion-1.1913401

National Youth Council of Ireland. 2015. "Living Wage Forum Welcome, but Increase in Temporary Contracts and Precarious Work Needs to Be Tackled." National Youth Council of Ireland, September 30 Retrieved from www.youth.ie/nyci/Living-Wage-forum-welcome-increase-temporary-contracts-and-precarious-work-needs-be-tackled

Negra, Diane. 2014. "Adjusting Men and Abiding Mammies: Gendering the Recession in Ireland." In *Masculinity and Irish Popular Culture: Tiger's Tales*. C. Holohan and T. Tracy, eds. London: Palgrave, pp. 223–237.

New York Times. 2014. "Ireland Bucks the Trend in Eurozone's Stalled Recovery." *New York Times*, September 1.

Oireachtas Library and Research Service. 2013. "Responding to Youth Unemployment in Europe, 2013." Retrieved from www.oireachtas.ie/parliament/media/housesoftheoireachtas/libraryresearch/spotlights/Responding_to_Youth_Unemployment_in_Europe.pdf

O'Leary, Eleanor and Negra, Diane. 2016. "Emigration, Return Migration and Surprise Homecomings in Post-Celtic Tiger Ireland." *Irish Studies Review*, 24(2): 217–141 Retrieved from https://doi.org/10.1080/09670882.2016.1147406

O'Toole, Fintan. 2014. "Quickly but Quietly, Ireland Is Disappearing Its Young People." *Irish Times*, September 2 Retrieved from www.irishtimes.com/news/social-affairs/quickly-but-quietly-ireland-is-disappearing-its-young-people-1.1914554?page=1

Rattner, Steven. 2014. "Inequality, Unbelievably, Gets Worse." *New York Times*, November 16. Retrieved from www.nytimes.com/2014/11/17/opinion/inequality-unbelievably-gets-worse.html

Rogers, Stephen. 2014. "Jobseekers Who Refuse Placement Lose Dole Payment." *Irish Examiner*, May 12. Retrieved from www.irishexaminer.com/ireland/jobseekers-who-refuse-placement-lose-dole-payment-268248.html

Statistica. 2014. "Youth Unemployment in EU Member States as of May 2014 (Seasonally Adjusted)." *Statistica: The Statistics Portal*. Retrieved from www.statista.com/statistics/266228/youth-unemployment-rate-in-eu-countries (accessed December 4, 2014).

Tierney, Ciaran. 2014. "Emigration Cuts Deep into County Galway GAA Clubs and Community." *Connacht Tribune*, August 21 Retrieved from http://connachttribune.ie/emigration-cuts-deep-galway-gaa-clubs-community

Waterford Whispers News. 2014a. "Emigration Will Be Offered as a Leaving Cert Subject in 2015," June 5. Retrieved from http://waterfordwhispersnews.com/2014/06/05/emigration-will-be-offered-as-leaving-cert-subject-in-2015

Waterford Whispers News. 2014b. "The Top Five Destinations You Just HAVE to Emigrate to Post-Leaving Cert" August 13. Retrieved from http://waterfordwhispersnews.com/2014/08/13/the-top-five-destinations-you-just-have-to-emigrate-to-post-leaving-cert

We're Not Leaving. 2013. "On the Dáil Queue." Facebook, October 20. Retrieved from www.facebook.com/werenotleaving/photos/pb.551038758278980.-2207520000.1417696474./552739824775540/?type=3&theater

We're Not Leaving Galway. 2014. "Brand Ireland" [Online Video]. Retrieved from www.youtube.com/watch?v=6qNy0wBfdQ8

9 Youth unemployment, neoliberal reforms, and emigration in West Africa

Ange Bergson Lendja Ngnemzué and Tamar Mayer

Africa has the youngest population in the world (World Bank 2016) and also a high rate of youth unemployment. In 2015, children under the age of 15 accounted for 41% of the population and those between 15 to 24 accounted for a further 19% (United Nations Department of Economic and Social Affairs [UNDESA] 2015: 7). Moreover, in sub-Saharan Africa (SSA), 70% of the population is below the age of 30 and more than 200 million people are between the ages of 15 and 24 (UNDESA 2015: 7). A publication co-sponsored by the Agence Française de Développement (AFD) and the World Bank states that "each year, between 2015 and 2035, there will be half a million more 15-year-olds than the year before" (AFD/World Bank 2014: 2).

West Africa is one of the subregions where the population will grow most: over the next 35 years, "West, Central, and East African countries will experience large youth population increases" (African Development Bank [AfDB] 2015: 117). The impact of such growth, with "youth bulges," especially in countries that have no economic security, can easily lead to social and political unrest and slow economic growth. Labor markets in developing countries in general, and in Africa in particular, are unable to absorb annually the hundreds of thousands of new job seekers, and this leads many to seek employment opportunities elsewhere. As the population of West Africa continues to grow, the region is experiencing a large intraregional migration flow: the region houses more than 7.5 million migrants who originated mostly from another West African country and account for almost 3% of the region's population. Labor migration in West Africa, which has intensified since the 1990s (Ndiaye and Robin 2010b: 50), has also, as Okome (2012) has argued, been the direct result of neoliberal reforms enacted at the turn of the 1980s.

In this chapter we analyze the situation of youth unemployment in West Africa by focusing not only on the overall trends of the phenomenon itself, but also on the internal migration patterns that have exacerbated it, especially since the neoliberal reforms of the mid-1980s. We will consider how the geographical features of West Africa, as well as its climate and poor literacy rates, have contributed to this pattern of intraregional migration. We will also suggest a possible relationship in West Africa between youth unemployment and the lack of international mobility. Not least among the concerns regarding the future of young people in West Africa is the changing nature of employment in the region; we discuss the ways in which informal labor and the concept of underemployment[1] have affected the growing youth population.

Trends and the rise of youth unemployment in Africa

As discussed in the introduction to this volume, fluctuations in youth unemployment have varied over time and geographic region. Table 9.1 breaks down the rate of youth labor force participation by individual regions, comparing the 1991 to the 2014 rates. It shows that east and south Asia, for example, have seen a significant decline in labor force participation, trends that are not observed in Africa. There, specifically in North Africa and SSA, the total rate of young people's participation in the labor market has not changed much in the last 25 years. And even though this might be good news, the reality is that approximately one in two young people in SSA is unemployed or underemployed, and gender imbalance in labor participation in these regions is glaring. Far fewer young women are employed, and their rate of participation had declined in 2014 from the previous rate in 1991.

Most of the regions in the world have experienced a decline in the rate of youth participation in the labor market. According to the International Labour Organization (ILO) (2018: 2), the number of unemployed young people (under the age of 25) globally stands at 13%, which is 4.3% higher than the adult rate. The lack of employment opportunities for young people remains a global challenge. The difficulty in understanding the employability of young people in the context of Africa, in general, and SSA, in particular, is two-fold. First, statistics are inadequate in capturing those who are gainfully employed versus those who are underemployed or partly employed; and second, it is hard to capture the numbers of

Table 9.1 Youth labor force participation rates, by region and sex, 1991 and 2014

Region	1991			2014		
	Total	*Male*	*Female*	*Total*	*Male*	*Female*
World	59.0	67.0	50.6	47.3	55.2	38.9
Developed Economies and European Union	55.6	58.7	52.4	47.4	49.1	45.5
Central and South-Eastern Europe (non-EU)	50.2	56.3	44.0	40.6	47.9	33.0
East Asia	75.7	74.9	76.6	55.0	57.0	52.9
South-East Asia and the Pacific	59.3	65.8	52.7	52.4	59.4	45.2
South Asia	52.2	70.4	32.5	39.5	55.2	22.6
Latin America and the Caribbean	55.5	71.3	39.6	52.5	62.1	42.6
Middle East	35.6	57.3	12.6	31.3	47.2	13.8
North Africa	37.0	51.8	21.5	33.7	47.2	19.7
Sub-Saharan Africa	54.3	58.6	50.1	54.3	56.6	52.1

Source: ILO (2015a: 9).

young people who are truly unemployed. In SSA, many young people are either underemployed or are employed in the informal economy; they experience part-time work, low income, low productivity rates, and an inadequate use of their skills; and often, in order to lift themselves out of poverty, they join those who are unemployed in search of more stable jobs, where their skills can be utilized. As Table 9.1 shows, about 46% of African young people from the SSA region are not considered participants in the labor market, although they most likely are either underemployed or employed in the informal sector of the economy.

Even though regional or continental statistics concerning youth unemployment are important, they often mask the disparities that exist within the continent (AfDB 2015: 121, figure 5.4). In 2000, the average rate of youth unemployment in North Africa (Libya, Algeria, Morocco, Tunisia, and Egypt) was about 34%, with Algeria leading at over 50%. By 2013, however, the same North African countries had an average youth unemployment rate above 40%, an increase from 2000 (with Morocco being the lowest at slightly less than 20% and Libya the highest with more than 50% of the youth population unemployed). During this period, Algeria's rate of youth unemployment had dropped dramatically by more than half, to about 24%. Similar disparities are found in other parts of Africa. In southern Africa, the average rate of youth unemployment was about 40%, while South Africa alone exhibited in 2000 a rate of 45%, but by 2013 had the highest youth unemployment rate in the region, at above 50% (AfDB 2015: 114–145).

Table 9.2 Youth unemployment and working poverty trends and projections (2015–2017)

	Youth Unemployment Rate, 2015–2017 (Percentage)			Unemployed Young People, 2015–2017 (Millions)		
	2015	*2016*	*2017*	*2015*	*2016*	*2017*
World	12.9	13.1	13.1	70.5	71.0	71.0
Developed Countries	15.0	14.5	14.3	10.2	9.8	9.6
Emerging Countries	13.3	13.6	13.7	52.9	53.5	53.5
Developing Countries	9.4	9.5	9.4	7.4	7.7	7.9

	Young People Working in Poverty 2015–2017 (Percentage)			Young People Working in Poverty 2015–2017 (Millions)		
	2015	*2016*	*2017*	*2015*	*2016*	*2017*
Total Emerging and Developing	38.4	37.7	36.9	159.9	156.0	152.2
Emerging Countries	31.2	38.2	29.3	107.3	102.7	98.4
Developing Countries	73.3	72.2	71.0	52.6	53.3	53.8

Source: ILO (2016b).

Youth unemployment rate in West Africa, by country (2000 and 2013)

As shown in this section, from 2000 to 2013 the unemployment rate in West African countries did not change. Table 9.2 shows that no West African country has a youth unemployment rate higher than 20%. The reality behind the statistics is far more complicated, however, especially since we know that northern Africa and SSA have shown slow progress in terms of the availability of jobs, their quality, and productivity (ILO 2016a). In the case of Benin, for example, where rates of youth unemployment are low, but so are the employment figures (AfDB 2015), a large share of the young population appears to be economically inactive. Most of these young people have had poor education opportunities and, therefore, as in other developing countries, participate only in the "low-skill" economy (AfDB 2015: 121). The case of Benin is particularly interesting and shows the problems that we face when we expect statistics to point to trends of employment/unemployment. Benin's actual unemployment rate is very low (0.3%) and figures offered in the World Bank report (2014: 7) indicate that rate to be at 4.9%. When we parcel the unemployed population we see, as the World Bank report shows, that these higher rates (4.9%) refer specifically to young people with higher education, who because they come from well-to-do families, can take the time and wait for the right job (World Bank 2014). Most people cannot wait for the right job or the right time; with low education and looming poverty they find their way to the informal economy and are underemployed, a category that appears in the statistics as "economically inactive." This problem is not unique to Benin and can be found in many developing countries.

When 90% of the global youth population lacks quality employment (ILO 2013: 1) and about 2.5 billion people, or half of the global workforce, is employed in the informal economy (ILO 2015b: 3), many of them (156 million or 37.7% of working young people) are relegated to extreme or moderate poverty (compared to 26% of working adults) (ILO 2016a). In West Africa, roughly one in seven young people (cf. the average of the figures in Table 9.3) is not working and about 38% of those are willing to migrate, which puts the migration rate of young people as the highest in the world (ILO 2016a).

Unemployment and migration

There are several local and regional factors that affect youth unemployment and migration in West Africa. These include illiteracy rates, climate change, and the physical terrain and road infrastructures in the region. We will discuss each of these in turn below.

West Africa leads the world in illiteracy rates (United Nations Development Programme [UNDP] 2011). According to Caroline Pearce (2009), who carried out an Oxfam study, 65 million West African adults, which is about 41% of the total population, can neither read nor write; this reduces significantly their employability and leads to a host of problems, including social and professional mobility (World Bank 2016), susceptibility to layoffs, and eventually social and economic exclusion. Of the 15 countries that make up West Africa, in 13 of them between half and three-quarters of the population is illiterate (see Table 9.3). With such high levels of illiteracy it is hardly possible to expect stable levels of employability (at the individual level) and economic growth (at the national level). This labor force in flux, then, must look for economic opportunities in other destinations.

The second predictor of increasing underemployment of West African young people is the harsh Sahelian climate, which greatly reduces the workday and therefore the productivity of economic agents. In many West African communities, people are forced to work only in the morning, since the suffocating heat of the afternoon (above 40°C) precludes any major physical labor or activity. For farmers this is a serious handicap, forcing entire agricultural communities to seek employment indoors and elsewhere. This has led to an increase in rural-to-urban migration, placing further pressures on cities. As global warming progresses, heat waves in the tropics will be harsher and more deadly (Mora et al. 2017), and this will lead the already marginal lands to be even drier and far less productive, forcing millions to move. Climatic migrations will challenge cities and countries, in the region and overseas, to absorb the newcomers and provide them with employment opportunities and housing. Such a feat may not be possible given

Table 9.3 Youth unemployment rate in West Africa, by country, 2000 and 2013

West African Country	2000				2013			
	1–5%	5–10%	10–15%	15–20%	1–5%	5–10%	10–15%	15–20%
Benin	X				X			
Guinea	X				X			
Liberia		X				X		
Burkina Faso		X				X		
Sierra Leone		X				X		
Côte d'Ivoire		X				X		
Niger		X				X		
Cabo Verde			X				X	
Mali			X				X	
Guinea Bissau			X				X	
Gambia, The			X				X	
Togo			X				X	
Nigeria			X				X	
Senegal				X				X
Ghana				X				X

Source: Lendja Ngnemzué, adapted from AfDB (2015: 121).

the internal migrations of young job seekers within these countries and the weak economic prospects of many of them.

The third factor that eases and maybe even encourages migration within the region (and may explain the fluctuation in employment figures) has historical and geographical roots. Regional history and political geography, as well as the informal roads infrastructure that has developed over time, may explain or even encourage young people to pick up and leave in search of work, instead of being steadfast and looking for work locally. The history of the region, dating back to the Great Sahelian empires (from the 9th to the 15th century) (Demba Fall 2004), has led to both linguistic and demographic connections among the many countries of West Africa as well as to ties among and between peoples and places in the region. Because 30% of West African languages are cross-border languages, much of the region is seen, at least ethnolinguistically, as a continuum (Demba Fall 2004: 3). This, together with the vast expanse of landscape created by the semi-arid climate, has resulted in fluid, unpoliced, and under-surveilled borders. Further, the treaty signed in Lagos, Nigeria, on May 28, 1975, affirms "the abolition as between the Member States of the obstacles to the free movements of persons, services and capital."[2] With "obstacles to the free movements of persons, services and capital" being removed, a relatively free migration of individuals seeking employment was enabled. The Economic Community of West African States (ECOWAS) Dakar Protocol,[3] signed in 1979, went even further, and pooled work opportunities in the region, legitimizing at the same time the sociodemographic community and "the will of individuals taken in isolation to achieve their purpose in spaces other than the nation-state" (Demba Fall 2004: 7).

High rates of illiteracy and physical conditions may have contributed to levels of unemployment, but infrastructure, as well as the historical patterns of regional movement reinforced by treaties and protocols, have made it easier to migrate for work. Neoliberal reforms and structural adjustment programs have worsened the employment opportunities at home, exacerbating the speed and volume of migration, and thus worsening the unemployment situation—creating uncertainty not only in places of origin but also in destinations as well.

From neoliberalism to declining social integration

Post-independence, specifically between 1960 and 1980, authoritarian neopatrimonial regimes[4] blossomed across SSA. Under their governance, public and private sectors were indistinguishable and the population in the region experienced full employment. Many of the countries in West Africa depended heavily on agriculture to boost their growth, and large segments of the population were employed in this sector of the economy. In Côte d'Ivoire, for example, in 1970, agriculture accounted for 90% of its exports (Decraene 1971: 17–18), providing jobs for all. Further, as some of these economies grew and somewhat diversified, "the hiring of young graduates was automatic and even a right for those who did not find [work] elsewhere" (Decraene 1971; Mbembe 1985; Médard 1995). In other countries, where liberal ideology prevailed, part of their economic plans included active recruitment of young workers. In post-colonial Africa, the early 1970s were characterized by insignificant unemployment rates (Charmes 1996: 499), a trend that would be reversed just a short decade later when neoliberal policies and structural adjustment programs were instituted.

As the value of agricultural products in the late 1970s experienced a 60% decline (Mbembe 1985) and a crisis in the state economy and structure ensued, massive unemployment of both untrained and well-trained young people followed (Mbembe 1985). In an effort to keep development on track and address the crisis, the World Bank and International Monetary Fund (IMF) adjusted the economy by imposing austerity plans that transferred social burdens from the state to individual households. By introducing economic deregulations and encouraging competition, and by restructuring the economy, these programs resulted in massive unemployment, on the one hand, and flooding of the informal economy, on the other. Neoliberal policies, here as elsewhere in Africa, had deepened poverty among young people, and these working poor, who came to be referred to as *flat feet*—those who would never be described using the English colloquial term of *well-heeled*—have become the archetypal victims of these reforms (Lendja Ngnemzué 2010: 152–153, 2016: 17–18).

"Flat feet" and the ascendancy of the informal sector

The relationships between neoliberal policies and youth unemployment have been discussed earlier in several of the chapters in this volume. In the case of most developing countries, structural adjustment programs and neoliberal policies, along with mechanization and industrialization (Charmes 2001: 243), unleashed a major rural-to-urban migration, and these *flat feet* who lack skills often find themselves scraping out a livelihood in the informal sector of the economy. The bigger the rural exodus, the larger the informal sector becomes, and the more intense the negative social, economic, and political ramifications.

While the informal sector may offer a vast array of employment opportunities, the precarious nature of the work and the associated uncertain income, which are typical of that sector's practices and standards, serve to the disadvantage of many young people. It keeps them poor, unskilled, and in a position that prevents them from achieving their full potential. Further, a large informal economy significantly reduces the capacity of political authorities to organize the labor market by setting common standards for all. The informal sector thereby weakens new players in the name of liberal flexibility, so that young people are forced to work in it without rights and any social security.

We suggest that the massive rural-to-urban migrations specifically of young people, and the abundant labor that they represent, have reinforced the precarity of these people's labor and enshrined the economy's productivity in informal practices. As a result, the number of poor people in SSA has grown by about 100 million in 25 years (1990–2015) "from 288 to 389 million" (Beegle et al. 2016: 4); in West Africa, 70% of young people are considered to be working poor. *Flat feet* not only epitomize a declining social integration—for they are often on the move—but also represent job insecurity in the context of neoliberal flexibility. They epitomize the dangerous trend in which the informal sector is becoming standardized, making insecure and temporary work the basic norm. Therefore, with neoliberal deregulations, having a job does not enable certain categories of workers to earn a living. Because work may be lost at any time or because they may not be paid for their work, those deemed to be *flat feet* are similar in some ways to those who are unemployed altogether (Agamben 1997).

Topology of destinations reconsidered

Some *flat feet* in dire situations, some literally for want of food, try to (re)make a living by emigration or to seek other forms of survival. Being considered in their own countries as a dangerous social category,[5] *flat feet* are now moving to the West. Such a move fits within the neoclassical economic theory of the relationship between unemployment and migration. Arthur Lewis (1954) stated that international migration is an ad hoc factor of the international production system, which acts as a simple adjustment variable in the regulation of the supply of (and demand for) labor. Traditional economies (in the case of West Africa) are affected by permanent underemployment due to the concentration of the labor force in a primary sector that lacks jobs and, therefore, in accordance with Lewis's theory, international migration could solve underemployment pressures. But this theory is simplistic and does not address the multiscalar realities that in later years were offered by critical sociologists. While these sociologists, particularly Frank (1970), Amin (1973, 1994), Frank and Amin (1978), Wallerstein (1990), and Sassen (2013), provide excellent explanations for current poverty and show the role of colonialism and imperialism in creating underdevelopment in Asia, Africa, and Latin America, their theories do not address the most important questions for the discussion of poverty and unemployment in the developing world: how to alleviate poverty, how to abolish under- or unemployment, and whether migration is indeed a solution.

An ILO report quoted by the United Nations (UN) in 2016 indicated that there are 11.3 million unemployed young people in SSA compared to 71 million in the world (UN 2016). The AfDB (2012) noted that, if Africa remains underdeveloped in 2040, most of the billion young people in the African labor market will be increasingly looking to migrate to rich countries. Echoing these predictions, a UN authority has noted that "the West African youth, disillusioned by mass unemployment, finds itself increasingly confronted with two choices: violence or migration, which in its turn represents a security risk for advanced democracies."[6] Reports in Western media declaring that migration from Africa is unstoppable, showing images of boats full to capacity with young Africans who are hoping to reach European shores (which sometimes capsize and sometimes do not), and relaying stories about West African migrants being bought and sold openly in slave markets in Libya (Graham-Harrison 2017) all suggest that Africans are so desperate that they will risk everything, even their lives, in the hopes of a better future. The move north, though, is the desire not only of the un- and underemployed, but of adolescents as well (Timera 2001: 37).

The desperation of young Africans and, as Esse Amouzou (2009) claims, their migration, "speaks for itself." When so many risk their lives and try to reach a global space that shuts them out and rejects them, it is important to reevaluate the relationship between the individual and the state and between African satellites and European metropoles, to use A.G. Frank's language. The escape north is an escape from poverty, but it is also a way to escape the "*control* of central authorities without even subverting or destabilizing it" (Bayart 1999: 117). The SSA state, then, becomes a "place of passage or transit," a hurdle that "citizens seek to pass over" (Mbembe 2010: 22) in order to achieve their goal.

Breakdown of intelligibility

Despite statistics that point to future trends in migration of (West) Africans to Europe, and despite alarmist warnings from politicians and lay people that African migration to Europe cannot be stopped, the migration story of West African young people, especially their places of destination, is nuanced and requires some elaboration. A quick examination of the figures suggests that the alarmist forecasts are unfounded. Young Africans, for the most part, prefer to stay on the continent, and there has been a significant decline in their move to Europe as the former colonizers imposed migration restrictions for nationals from their previous colonial states (Flahaux and De Haas 2016). Of the 51 million international migrants aged 15–29, more than half live in developed countries. International migration today affects only 3.3% of the world population, although the phenomenon has intensified significantly in the first 15 years of the 21st century: international migration increased from 173 million in 2000 to 244 million in 2015 (UN 2016: 5) and the rate of migration has increased from an annual growth rate of 2% in the period 2000–2005 and more than 3% annually in the period 2005–2010 (UN 2016: 5). Since then, and for the period 2010–2015, however, the rate of migration to the West has declined and stands now at about 1.9% annually. Of the millions of international migrants to the West, Africans account for about 6 million (UN 2016: 5), or less than 0.5% of the total migration to the West in 2015 (UN 2016: 5). Further, of all the African migrants, particularly to Europe, West Africans constitute the second largest group of migrants, after North Africans and ahead of East and Central Africa (Flahaux and De Haas 2016: figure 9).

There are a few observations about (West) African migration that are worth noting. First, Africa seems to follow some of the general trends of international migration, at least as far as the median age of migrants is concerned. The median age of international migrants in 2017 was 39.2 years, compared with 38.0 years in 2000 (UN 2017: 16). This slight increase is seen in Africa as well, where the median age of African international migrants was 27 in 2000, 29 in 2015 (UN 2016: 1), and 30.9 in 2017 (UN 2017: 16). Even though we consider the youth category to be defined as those aged 15–24, people under the age of 20 are often underrepresented among international migrants. According to an UNDESA report between 2000 and 2015: "Globally, 15 per cent of all migrants were under 20 years of age, compared to 34 per cent of the total population" (2016: 12). Second, Africa sends the smallest numbers of migrants to the West. Of the 244 million international migrants to the West in 2015, 43% were born in Asia, 25% in Europe, 15% in Latin America and the Caribbean, and only 14% in Africa (UN 2016: 1). As immigration into Europe has become increasingly more difficult, there has been an increase in African migrants to other destinations like North America, India, Russia, Japan, Brazil, and Argentina (Henao 2009; in Flahaux and De Haas 2016: 14–23). Neither these rates nor the volume of migration support the alarmist reports of the increase in young African immigrants into Europe. Third, while Africans migrate in relatively small numbers to the West, and this is specifically true for those from SSA,[7] the larger share of their migration is within the continent and, specifically, within the regions of origin. UN figures for 2015 show that 66% of African migrants do not leave the continent; 70% in West Africa, 65% in Southern Africa, 50% in Central Africa, and 47% in East Africa migrate within the subregion (United Nations Economic Commission for Africa (UNECA/African Union 2016: 3). Only from North Africa do 90% of emigrants go to destinations outside of Africa, predominantly Europe, the Middle East, and North America (UNECA/African Union 2016: 3).

The intraregional migration in West Africa is to be expected given what we have discussed about *flat feet* and their search for work. The regional migration patterns facilitated by historical routes and treaties, signed in the latter part of the 20th century, have eased movement of people (and capital) across the region. West African young people roam large areas, following paths that have been renewed and reconstituted over the years, taking advantage also of the ethnolinguistic connections established centuries ago. They reach their destinations on foot, on the backs of donkeys, or by transportation such as buses or trucks. Only a small minority use either the old trans-Saharan routes or new ones that have been created by smugglers. While migration has been eased because of historical connections within the region, the flow is not assured in perpetuity and sometimes can come to a halt altogether and prevent migrants from reaching their European "El Dorado." This can happen at the place of destination or along the way, as states may impose their own immigration restrictions.

In recent years, for instance, countries such as Equatorial Guinea, Chad, and Angola have developed restrictive policies ranging from mass expulsions of foreigners, reduction of immigrants' rights, violation of subregional agreements, or nonratification of legal texts relating to the free movement of goods and persons. After the riots of May 2008, an outbreak of violence among part of the South African population against one sector, referred to by a term in the Zulu language that literally means "barbarians," is an emblematic case in this radical change in public immigration policies in Africa. In fact, this 2008 event kicked off the beginning of restrictive government policies on immigration from Africa. While the proliferation of expulsions and attacks against foreigners has led to a hardening of the positions of politicians and analysts on the "software" of immigration, these growing anti-immigrant riots do not reflect a structural xenophobia of the South African society as such; rather they plunge us into "the context of the post-apartheid management of migratory flows and that of a low-capacity environment of the State, [a] crisis of political succession and widening of socio-economic disparities" (Kabwe-Segatti 2008: 100).

Interruptions in migration flows and shattered dreams were particularly noted in 2011 during the "Arab Spring" and since 2015 in South Africa, where xenophobia has resulted in significant return-migration of West African young people, now seen as the real migration crisis within Africa. Thousands of African migrants are now forced to return to their home countries, because of political turmoil and lack of security in their countries of residence. These return migrants further aggravate the already strained and challenged labor markets, leading to further insecurity, because of the relentless pressures on resources and employment opportunities that their home countries have not been able to address in the first place.

Conclusion

Africa is facing a crisis: the population is very young, poor, and mostly uneducated, and its economic prospects are limited and often nonexistent. As a result, people are on the move hoping for a better future. If, in the first post-colonial period, 1960–1980, unemployment was a nonissue and underemployment did not really exist, the period after 1980 is characterized by massive unemployment and all the ills that accompany it. Pressures from the IMF and the World Bank to economically develop, as well as the ensuing structural adjustment of the economy, resulted in massive unemployment, which led to rural-to-urban migration; this led to social disintegration and the unmaking of many communities, resulting not only in the loss of a social safety net but also a life of poverty. With few prospects for a better future, many young Africans have resorted to migration.

Neoclassical economic theories predict that high unemployment leads to an increase in migration. Africa, therefore, is the quintessential example of the validity of such theories, for as countries experience economic crisis, emigration intensifies. Whereas, in the past, many used to immigrate to Europe, this trend has changed in the last two or three decades because European countries changed their immigration policies and closed their borders to migrants from their colonial states. Instead, Africans have resorted to migration within their continent and, specifically, intraregionally. (This is not to say that there are not thousands of young Africans, mostly men, who risk their lives in the hopes of reaching the West and whose images on packed boats in the Mediterranean feed the reports about massive African immigration to Europe. But as this chapter has shown, those who seek to immigrate to the West are a minority.) The case of West Africa, as we have illustrated, is particularly interesting because these migration patterns were established as early as the 9th century, when the region was part of the Greater Sahelian empires and people moved freely from place to place, creating connections among those within the region and to the region itself. These movements enabled different ethnic groups to settle in different locales across the region. The current migration patterns then rely on the old road system but also on the links among ethnic groups across borders. These communities ease the migration process. Moreover, because of its past, and certainly because of treaties signed among ECOWAS, which remove obstacles for the movement of people and capital, West Africa has become the most open region in Africa, enabling free migration within this region. Thus the *flat feet* who lack the resources needed to move to the West take advantage of the openness of ECOWAS.

The ability to move freely within their region and, to a lesser extent, within their continent, allows many young migrants to avoid risks they may encounter along the route to Europe (on land in Libya or in the Mediterranean). At the same time, however, their intraregional migration exacerbates an already delicate labor market in urban areas, often leading to violence and social disorder. Despite the tendency of international institutions and African governments to cope with the hegemony of the informal sector, each country cannot solve the problem alone. African young people are waiting for a future beyond that "*débrouille*": African countries must work on a political solution to address the frustration of *flat feet*.

Notes

1 Underemployment occurs when an employee works involuntarily in a part-time job, where their (higher-level) skills or education do not match the job, and where people work in the informal sector of the economy.
2 The treaty of ECOWAS was concluded at Lagos on May 28, 1975. Authentic texts in English and French were registered by Nigeria on June 28, 1976, article 2 (d) p. 20. Note re: Member States: Cape Verde has been a member of ECOWAS since 1976, and Mauritania left the organization in December 2000, while Morocco's petition to become the 16th member country in December 2017 has been put on hold at the time of writing.
3 The "Protocol of the Economic Community of West African States" (ECOWAS) on the free movement of persons, right of residence and establishment" (Protocol A/P1/5/79) adopted in Dakar on May 29, 1979, is among the most important ones which organize the modalities for the effective implementation of this free movement within the ECOWAS area. Three rights are guaranteed to the citizens of ECOWAS by this Treaty: (1) the right of entry and residence without prior authorization; (2) residence and work in all ECOWAS member countries; and (3) the right of establishment to practice liberal professions (doctors, lawyers, architects, etc.). Since it went into effect in 1984, the ECOWAS Protocol has asserted the principle of total free movement and does not provide for "enhanced protection at the external borders of outer space" as is the case in the Schengen Agreements (1985) that applied to the ten member states of the then European Economic Community (EEC).
4 Neopatrimonial regimes refer to a state structure whereby the regime uses state resources as a way to ensure the loyalty of their populations.
5 Paul Richards (1996) has analyzed the "crisis arc in West Africa" (Sierra Leone, Nigeria, and Liberia), attributing it to the neglect of young people by the political authorities of the region where the state is constructed, and existing in the midst of hereditary societies or residues of kingdoms that are hostile to the state. Michel Galy (2004) has already applied this hypothesis in his analysis of the Ivorian civil war (2002–2011), where young people were major players in violence either for or against the state. Biaya (2000: 14) considers that there are many enrolled young people in the

ranks of Boko Haram, in North Cameroon or Nigeria, getting involved as child soldiers, terrorists, or logistics staff, and girls, as objects of sex and rituals.

6 The Special Representative of the UN Secretary General for West Africa, during the presentation of the UN report on "Youth Unemployment and Regional Insecurity," Dakar, May 1999.

7 Ndiaye and Robin (2010a: 9) state that: "in Spain, one of the main immigration countries of the European Union, of six million people born abroad, only 3% are from sub-Saharan Africa against 41% from Europe and 33% from south America. In fact, 80% of sub-Saharans residing in Spain were born in one of the member states of the Economic Community of West African States."

References

African Development Bank (AfDB). 2012. "Labour Force Data Analysis: Guidelines with African Specificities." Retrieved from www.afdb.org/fileadmin/uploads/afdb/Documents/Publications/Labour%20Force%20Data%20Analysis_WEB.pdf

African Development Bank (AfDB). 2015. "Chapter 5: Africa's Youth in the Labour Market." In *African Development Report*, pp. 114–145. Retrieved from www.afdb.org/fileadmin/uploads/afdb/Documents/Publications/ADR15_UK.pdf

Agamben, Giorgio. 1997. *Homo sacer. Le pouvoir souverain et la vie nue* [*Homo Sacer. Sovereign Power and Bare Life*]. Paris: Seuil.

Agence Française de Développement (AFD)/World Bank. 2014. *L'emploi des jeunes en Afrique subsaharienne* [*Youth Employment in Sub-Saharan Africa*]. Washington, DC: International Bank for Reconstruction and Development/World Bank.

Amin, Samir. 1973. *Le développement inégal: Essai sur les formations sociales du capitalisme périphérique* [*Unequal Development: Essay on the Social Formations of Peripheral Capitalism*]. Paris: Editions de minuit.

Amin, Samir. 1994. "La nouvelle mondialisation capitaliste. Problèmes et perspectives" ["The New Capitalist Globalization. Problems and Perspectives"]. *Alternatives Sud*, 1(1): 19–44.

Amouzou, Essè. 2009. *Pauvreté, chômage et émigration des jeunes Africains: Quelles alternatives?* [*Poverty, Unemployment and Emigration of Young Africans: What Are the Alternatives?*] Paris: Editions L'Harmattan.

Bayart, Jean-François. 1999. "L'Afrique dans le monde: une histoire d'extraversion" ["Africa in the World: A Story of Extroversion"]. *Critique internationale*, 5(1): 97–120.

Beegle, Kathleen, Luc Christiaensen, Andrew Dabalen, and Isis Gaddis. 2016. *Poverty in a Rising Africa*. Washington, DC: World Bank.

Biaya, Tshikala K. 2000. "Jeunes et culture de la rue en Afrique urbaine" ["Youth and Street Culture in Urban Africa"]. *Politique africaine*, 4: 12–31.

Charmes, Jacques. 1996. "Emploi, informalisation, marginalisation? L'Afrique dans la crise et sous l'ajustement, 1975–1995" ["Employment, Informalization, Marginalization? Africa in Crisis and Under Adjustment, 1975–1995"]. In *Crise et population en Afrique, Crises économiques, programmes d'ajustement et dynamiques démographiques*, J. Coussy and J. Vallin, eds. Paris: CEPED, pp. 495–520.

Charmes, Jacques. 2001. "Flexibilité du travail, pluralité des normes, accumulation du capital économique et du capital social" ["Flexibility of Work, Plurality of Standards, Accumulation of Economic and Social Capital"]. In *Inégalités et politiques publiques en Afrique*. G. Winter, ed. Paris: Karthala-IRD Éditions, pp. 243–262.

Decraene, Philippe. 1971. "Côte-d'Ivoire: l'agriculture, pilier du développement économique" ["Ivory Coast: Agriculture, a Pillar of Economic Development"]. *Le Monde diplomatique*, February: 17–18.

Demba Fall, Papa. 2004. *État-nation et migrations en Afrique de l'ouest: le défi de la mondialisation* [*Nationhood and Migration in West Africa: The Challenge of Globalization*]. Paris: UNESCO.

Flahaux, Marie-Laurence, and Hein De Haas. 2016. "African Migration: Trends, Patterns, Drivers." *Comparative Migration Studies*, 4: 1. https://doi.org/10.1186/s40878-015-0015-6.

Frank, André Gunder. 1970. "Development of Underdevelopment." In *Imperialism and Underdevelopment: A Reader*. R.I. Rhodes, ed. New York: Monthly Review Press, pp. 4–17.

Frank, André Gunder, and Samir Amin. 1978. *L'accumulation dépendante* [*Dependent Accumulation*]. Paris: Éditions Anthropos.

Galy, Michel. 2004. "De la guerre nomade: sept approches du conflit autour de la Côte d'Ivoire" ["The Nomad War: Seven Approaches to the Conflict along the Ivory Coast"]. *Cultures & Conflicts*, 55: 163–196.

Graham-Harrison, Emma. 2017. "Migrants from West Africa Being Sold in Libyan Slave Markets." *Guardian*, April 10.

Henao, Luis Andres. 2009. "African Immigrants Drift toward Latin America." *Reuters*, November 15.

International Labour Organization (ILO). 2013. "Global Employment Trends for Youth 2013: A Generation at Risk, Executive Summary." Retrieved from www.ilo.org/wcmsp5/groups/public/—dgreports/—dcomm/documents/publication/wcms_212899.pdf

International Labour Organization (ILO). 2015a. *Global Employment: Trends for Youth 2015. Scaling Up Investments in Decent Jobs for Youth*. Retrieved from http://www.ilo.org/wcmsp5/groups/public/—dgreports/—dcomm/—publ/documents/publication/wcms_412015.pdf

International Labour Organization (ILO). 2015b. "Transition from the Informal to the Formal Economy Recommendation (No. 204): Workers' Guide." Retrieved from www.ilo.org/wcmsp5/groups/public/—ed_dialogue/—actrav/documents/publication/wcms_545928.pdf

International Labour Organization (ILO). 2016a. "Facing the Growing Unemployment Challenges in Africa." Press release, January 20. Retrieved from www.ilo.org/addisababa/media-centre/pr/WCMS_444474/lang—en/index.htm

International Labour Organization (ILO). 2016b. "World Employment and Social Outlook 2016: Trends for Youth 'Global Youth Unemployment Is on the Rise Again.'" Press release, August 24. Retrieved from www.ilo.org/global/about-the-ilo/newsroom/news/WCMS_513728/lang—en/index.htm

International Labour Organization (ILO). 2018. "World Employment Social Outlook: Trends 2018." Retrieved from www.ilo.org/wcmsp5/groups/public/—dgreports/—dcomm/—publ/documents/publication/wcms_615594.pdf

Kabwe-Segatti, Aurelia Wa. 2008. "Violences xénophobes en Afrique du Sud: retour sur un désastre annoncé" ["Xenophobic Violence in South Africa: Return of an Announced Disaster"]. *Politique africaine*, 4(112): 99–118.

Lendja Ngnemzué, Ange Bergson. 2010. *Politique et émigration irrégulière en Afrique. Enjeux d'une débrouille par temps de crise* [*Politics and Irregular Emigration in Africa. The Challenges of Struggling in Times of Crisis*]. Paris: Karthala Éditions.

Lendja Ngnemzué, Ange Bergson. 2016. *Gouverner et faire partir. Une sociologie des mobilités africaines par effraction vers l'Europe* [*Govern and Let Go. A Sociology of*

African Mobilities Breaking into Europe]. Vol. 2 of the diploma of d'Habilitation à Diriger des Recherches (HDR) in sociology. Dissertation, Ecole des Hautes Etudes en Sciences Sociales (EHESS), under the supervision of Rémy Bazenguissa-Ganga, Director of Studies, EHESS, Paris, December.

Lewis, Arthur. 1954. "Economic Development with Unlimited Supplies of Labor." *The Manchester School of Economic and Social Studies*, 22: 139–191.

Mbembe, Achille. 1985. *Les jeunes et l'ordre politique en Afrique noire* [*Young People and the Political Order in Black Africa*]. Paris: Éditions L'Harmattan.

Mbembe, Achille. 2010. *Sortir de la grande nuit* [*Leaving the Big Night*]. Paris: La découverte.

Médard, Jean-François. 1995. "État, démocratie et développement: L'expérience du Cameroun" ["State, Democracy and Development: The Cameroon Experience"]. In *Développer par la démocratie?* S. Mappa, ed. Paris: Forum de Delphes, Karthala, pp. 355–390.

Mora, Camilo, Bénédicte Dousset, Iain R. Caldwell, et al. 2017. "Global Risk of Deadly Heat." *Nature Climate Change*, 7(7): 501–506.

Ndiaye, Mandiogou, and Nelly Robin. 2010a. *Les migrations internationales en Afrique de l'Ouest: Une dynamique de régionalisation articulée à la mondialisation* [*International Migration in West Africa: A Regionalization Dynamic Linked to Globalization*]. IMI Working Paper 23. Oxford: International Migration Institute, University of Oxford.

Ndiaye, Mandiogou, and Nelly Robin. 2010b. "Les migrations internationales en Afrique de l'Ouest. Une dynamique de régionalisation renouvelée. Hommes et migrations" ["International Migration in West Africa. A Renewed Regionalization Dynamic. Men and Migration"]. *Revue française de référence sur les dynamiques migratoires* (1286–1287): 48–61.

Okome, Mojubaolu ed. 2012. *West African Migrations: Transnational and Global Pathways in a New Century.* Basingstoke: Palgrave Macmillan.

Pearce, Caroline. 2009. "From Closed Books to Open Doors: West Africa's Literacy Challenge." Retrieved from https://policy-practice.oxfam.org.uk/publications/from-closed-books-to-open-doors-west-africas-literacy-challenge-112394

Richards, Paul. 1996. *Fighting for the Rain Forest*. Oxford: James Currey.

Sassen, Saskia, ed. 2013. *Deciphering the Global: Its Scales, Spaces and Subjects*. New York and London: Routledge.

Timera, Mahamet. 2001. "Les migrations des jeunes Sahéliens: affirmation de soi et émancipation" ["The Migration of Young Sahelians: Self-Assertion and Emancipation"]. *Autrepart*, 2: 37–49.

United Nations (UN). 2016. "Global Youth Unemployment Is on the Rise Again." Office of the Secretary-General's Envoy on Youth, August 8. Retrieved from www.un.org/youthenvoy/2016/08/global-youth-unemployment-rise

United Nations (UN). 2017. "International Migration Report, Highlights." Retrieved from www.un.org/development/desa/publications/international-migration-report-2017.html

United Nations Department of Economic and Social Affairs (UNDESA). 2015. *Population Division 7 World Population Prospects: The 2015 Revision, Key Findings and Advance Tables, 2015*. New York: United Nations.

United Nations Department of Economic and Social Affairs (UNDESA). 2016. *International Migration Report, 2015, Highlights*. New York: United Nations.

United Nations Development Programme (UNDP). 2011. *Human Development Report 2011. Sustainability and Equity: A Better Future for All*. Human Development Index (HDI). New York: UNDP.

United Nations Economic Commission for Africa (UNECA)/African Union. 2016. "Side Event—International Migration in Africa: Framing the Issues." An Issues Paper. Addis Ababa: UNECA/African Union, March 25.

Wallerstein, Immanuel. 1990. "L'Occident, le capitalisme et le système-monde moderne" ["The West, Capitalism and the Modern World System"]. *Sociologie et sociétés*, 22(1): 15–52.

World Bank. 2014. "Benin—Youth Employment Project," February 13. Retrieved from http://documents.worldbank.org/curated/en/459301468204890551/Benin-Youth-Employ ment-Project

World Bank. 2016. "World Development Indicators." Washington, DC: World Bank. Retrieved from http://databank.worldbank.org/data/download/site-content/wdi-2016-highlights-featuring-sdgs-booklet.pdf

10 Sitting amid a pile of jewels

Youth unemployment and waste recycling in China

Carlo Inverardi-Ferri

> People walking by would often see him sitting happily in the middle of these high-quality scraps, looking as if he were sitting amid a pile of jewels.
>
> (Yu 2009: 390)

In his coming-of-age novel *Brothers*, the Chinese writer Yu Hua describes the history of his country from the years of the Cultural Revolution to the emergence of capitalism in contemporary China, narrating the story of two stepbrothers. One is Song Gang, a worker in a state-owned company, who is laid off and falls into disgrace. The other is Baldy Li, a scrap collector, who builds a millionaire empire peddling used Japanese suits (as the epigraph taken from Yu Hua's novel implies). Youth unemployment and waste recycling are a central part of Chinese society today and provide a meaningful perspective for our understanding of the social and economic changes of the country. They also play, not surprisingly, a fascinating role in literature and popular culture. In this chapter I look at the relationship between waste recycling and unemployment as well as the various countertactics that young people adopt to make a living and improve their material conditions in Chinese cities.

The emergence of youth unemployment in China has been the unpleasant but necessary by-product of the shift in its political economy. In little less than four decades China has moved from being a country deeply grounded in socialist structures to one of the fastest-growing market economies in the world. While cheap labor has attracted foreign investments in the aftermath of the launch of the reform and opening up policies, China has progressively moved up the global value chains to become the largest exporter on the planet, a nuclear power, and an influential geopolitical actor. Chinese companies are now integrated in global production networks (Coe and Yeung 2015; Yeung and Olds 2000) and have plunged into international financial markets (Wójcik and Camilleri 2015); however, this great transformation (Hurst 2009; Polanyi 1944) has not occurred without major compromises. As William Hurst (2009: 1) aptly points out, these "impressive gains brought about significant social dislocation, in particular for groups that had been winners under socialism, but found themselves losers in the new post-socialist order." The implementation of restructuring policies, decollectivization of land, and the abolition of communal production in the countryside have contributed to hasten this process (Hurst 2009) and made visible the rural surplus population (Fan 2008), which was previously hidden in the phenomenon of underemployment (Lee and Warner 2007).

Pushed to leave the countryside by raising unemployment rates, many young migrants moved to urban areas. While some certainly contributed to broaden the ranks of the urban proletariat, others turned to the informal sector (also called labor informality or the gray economy) as a way to make a living (Qian 2015). These included a variety of practices such as the hawker trade, motorcycle taxis, and waste recycling. This phenomenon is not peculiar to China but common to many economies in transition (Jeffrey and Dyson 2013). Yet, as Jeffrey points out, geographical research has paid little attention to the connection between youth unemployment and informality, and in particular to the question of how forms of fall-back work contribute to shaping different labor regimes (Herod 1997; Jeffrey 2008). This chapter thus attempts to fill this gap by analyzing the practices of young waste recyclers in Beijing, a group mostly composed of migrant workers from rural regions. It suggests that, for many young migrants, turning to informality is a conscious strategy to make a living and gain control over the amount of time they engage in labor.

My research for this chapter builds upon long-term ethnographic fieldwork among waste recyclers in Beijing. It is worth mentioning that almost every informant cited in this chapter is male. This reflects the fact that males are predominant in the industry. Here I acknowledge my own position as a Western man. Participant observation and semi-structured interviews were not only conducted independently by myself, but also collaboratively with Chinese scholars. This enabled the triangulation of sources. The remainder of this chapter is organized in three parts. I first delve into the territories of geographical literature on youth unemployment and labor informality, then move on to describe empirical materials collected in China. In conclusion, I assess the wider implication of analyzing unemployment through the perspective of informality as a key element of the political economy of capitalism.

Youth, informality, and survival practices

Research on youth has a central place in the social sciences today (Cheng 2016; Evans 2008; Skelton 2009). The investigation of processes associated with young people has identified themes and concepts central to geography (Jeffrey 2010, 2012, 2013). By investigating the difficulties young people encounter in finding employment, scholars have given voice to the hardship that many experience in work settings (Jeffrey 2010). While geographers studying childhood have focused on paid labor and domestic work, researchers of the older youth category have paid particular attention to the hardships young people encounter in their passage to adulthood. It is argued that global economic transformations have resulted in dramatic consequences for young people, who nowadays experience their passage to adulthood while being entangled in both global and local dynamics that previous generations have not encountered (Jeffrey and Mcdowell 2004).

Transition to adulthood for many young women and men has become a complex process where everyday experiences have to be negotiated in the context of wide structural phenomena such as harsh market competition, deep economic restructuring, and changing political orders, to cite just a few factors. The state, in both of the areas usually labeled as the global North and the global South, has increasingly retreated from the field of welfare, while the costs of social reproduction have progressively shifted to households and individuals, making young people more subjected to processes of isolation and marginalization (Jeffrey and Mcdowell 2004).

In China, while reform and opening up policies have brought about economic growth and new remunerative opportunities (Zhang 2000), the entire population has not equally benefited from the marketization process. Instead, new patterns of uneven development have emerged within the country engendering different forms of inequalities (Hudson 2016). As pointed out by Hurst (2009), new groups of dispossessed and marginalized have appeared in the aftermath of the reforms that, besides development and modernity, also brought about social dislocation and unemployment. As a result, in post-Maoist China some groups were able to benefit from new employment opportunities while others were forced to adopt a variety of strategies for getting by in their struggle for a better life. In this regard, the informal economy has served as a safety net for many young migrants who find in informality a way to make a living in the city. Various forms of informal labor have appeared in urban China and this chapter analyzes some of them, showing, at the same time, how young migrants consciously and unconsciously act as elements of change. While operating informally, young migrants shape the spaces of their cities in particular new ways. The analysis of how these spaces come into being can greatly contribute to our understanding of the unemployment-informality nexus (Jeffrey 2008), broadening our knowledge of economic geography, both empirically and theoretically (Yeung and Lin 2003).

As pointed out by Schucher, research on China has produced a scattered literature, only marginally investigating the theme of youth unemployment and the connection with informal labor (Schucher 2014, 2015). Research has paid particular attention to the theme of laid-off workers (Gold et al. 2009; Hurst 2009) and, to a minor extent, to unemployment in the countryside (Murphy and Tao 2007; Webber and Zhu 2007). More recent literature has investigated the rising phenomenon of graduate unemployment (Bai 2006). Only a few studies, however, have focused on atypical work for young people (Smith and Chan 2015), paying even less attention to informal practices (Schucher 2015).

My goal in this chapter is thus to shed light on the daily strategies of young migrants in Beijing and to analyze how their everyday practices serve both as a safety net and as an element of heterogeneity in the Chinese capitalist mode of production. Therefore, in this chapter I bring literature on youth unemployment into the conversation with scholarship on the political economy of capitalism to analyze the link between unemployment and informality in China, and to show how informal practices are a powerful means to provide an increase in real incomes for the unemployed and the dispossessed (Meagher 2013). As pointed out by Ferguson (2013: 231), informality is "less about producing goods and services than it is about finding opportunities [...] Such activities facilitate a kind of day-to-day survival on the part of the unemployed." While engrained into the overall Chinese economy, informality is an element of diversity and, as I suggest, a moment of resilience to capitalist social relations.

In the next section I will introduce a very brief historical vignette of recycling in China. My purpose is to give a few but nevertheless important insights on the different geographies of waste produced at specific moments in the history of the country. Then I turn to an empirical focus that provides an analysis of the political economy of youth unemployment and informality in contemporary urban China.

Youth unemployment and waste recycling in contemporary China

Rather than being a mundane concern of environmental governance, waste is an integral element of political economy. In China, the management and recycling of everyday refuse is a process that has significantly changed with the different political economies of the Republican, Socialist, and post-Maoist eras. These metamorphoses in time affected absolute, relative, and relational spaces (Harvey 2006; Lefebvre 1991)—in other words, the different dimensions of production (destruction) and exchange as well as the social relations that underpinned processes of waste management. I further explore in the following pages the rise of these variegated geographies of waste (Inverardi-Ferri 2017b, 2018). Here I show how the production of spaces of waste is a process unevenly intertwined with the phenomena of youth unemployment and informality.

Waste recycling has a long history in China. A typical figure of the geography of the Republican era (1911–1937) was the street collector or ragpicker, one of the main agents of the process of anchoring the past into the present (Dong 2003). As suggested by Goldstein (2006), in that era very little was wasted, but salvaged commodities were carefully sorted and promptly reused. Refuse passed through the hands of workers, merchants, and consumers, to be reinvented in their meanings and material lives (Goldstein 2006). Several "dark markets" appeared in Beijing at the time. Here it was possible to meet merchants, artists, and entertainers and buy all sorts of recyclable goods (Dong 1999). Some of these markets appeared in the small hours of the night only to disappear in the early morning. Others were tolerated and became an integral part of Beijing. The biggest of these places was Tianqiao, a dark market situated in the southern part of the city, under the shadow of the Temple of Heaven (Dong 1999, 2003).

In 1949, with the advent of the Communist Party and the reorganization of the social, economic, and spatial life in Maoist China (1949–1978), imaginaries of waste and conceptions of waste management were subjected to major changes (Ensmenger et al. 2005). In Beijing, a new system for waste sorting was gradually implemented during the 1950s. Recyclers were transformed into state workers, while dark markets for recyclables disappeared from the map of Chinese cities (Goldstein 2006). In an ironic twist of history, however, ragpickers and hawkers, who had characterized the Republican era, were destined to reappear in Chinese cities in the 1980s, to personify, once again, the paradoxes of development.

From the 1980s, consecutive waves of workers moved from rural regions into coastal metropolises, attracted by the potential to improve their material conditions. In the cities, young migrants became associated with the humblest activities. Among these, waste recycling and the trade of secondhand goods played a major role in absorbing the displaced rural working force (Kirby and Lora-Wainwright 2015). The return of waste pickers in contemporary China is therefore engrained in a number of major shifts that affected the country, such as the marketization and monetization of the economy, the changes in land policies and markets, the emergence of narratives on environmental protection and sustainable development, and the process of internal migration (Goldstein 2006) driven by the emergence of unemployment in rural China.

The following pages particularly focus on the connection between unemployment and waste recycling, and how they affected the geographies of Chinese cities. In the last three decades, Chinese cities have experienced huge waves of urbanization. This process has changed the physical landscape of metropolises and had major consequences in reworking imaginaries around waste and recycling in the country. The urbanization process has had a symbiotic relationship with the emergence of informal recycling. Indeed, the very collapse of the system constructed under the planned economy was driven by the physical transformation of urban centers. In a society that had increasingly adopted new behaviors of consumption and disposability, the recycling network inherited from the socialist era became itself a disposable vestige (Minter 2013). In the remaking of the urban space, official planning simply erased from new master plans the collection places and treatment plants for recyclables, which gradually disappeared from the map of Chinese cities as urban redevelopment projects increased (Ensmenger et al. 2005). This particular conjuncture made possible a new geography of recycling, originally composed of small individual and family businesses, which came to light in many Chinese coastal metropolises. As explained in the account below, young migrant workers from poor rural regions found in the trading of recyclables an opportunity to improve their material means. Lao Wang,[1] a local informant, explains that waste recycling is a way for many who possess the qualities of strength and modesty to get out of the poor environment of their home region:

> Xinyang is quite far from the centre of Henan, so it rarely receives attention from authorities or supporting policies, public employment is rare. Moreover, the counties near the Dabie Mountains such as Guangshan, Luoshan, Xinxian or Jianjun are the poorest in the whole of China. So, poverty and lack of work pushed us to leave the region to look for opportunities elsewhere.

The account above highlights how many young migrants, pushed by unemployment in their home region, move to Beijing looking for job opportunities; however, while some enter the ranks of the urban proletariat through waged employment, others turn to informal labor. A number of reasons motivate this decision. As one informant explained to me in a personal communication, for migrants seeking their fortune in Beijing, their poor background of origin strengthens the qualities that help them to endure the hardships of life, making it easier for them to embrace stigmatized labor activities, such as waste recycling, which local urban citizens avoid.

Furthermore, starting a business in the waste industry enables young migrants to experience "more freedom," compared to waged work, and become independent "bosses." Turning to informality therefore represents a strategy to gain control over their labor time and escape the discipline of the clock experienced under waged work. It is therefore a process that enables worker agency (Herod 1997). Through a number of small acts that ensure they are economically getting by, these young migrants are able to engage in social relations that they consider to be less coercive (Katz 2004; Coe 2013; Coe and Jordhus-Lier 2011). This was the case for Lao Zhang, a young migrant who arrived in Beijing in his early 20s. After a short period working for a construction company he decided to become a scrap trader, judging that the business of recycling was a better way, for him, to make a living.

The choice of Lao Zhang is not an isolated one. Every day, many young migrants like him move to Beijing from poorer regions in the quest for a better life. The youngest of them often come alone, leaving their family in their home region. If already married, they send money home to support their wives and children, and they go back to visit them every few months. These young men find support in fellow villagers in the metropolis, where kinship networks serve as social glue in the local migrant community, providing help to newcomers. Migrants often live together in large informal settlements on the outskirts of the city, where a bed in a big dormitory can be rented for less than 10 Yuan. Unsurprisingly, these sites are extremely gendered places, where the majority of the population is male. Living and health conditions are often very poor. The coming of age of these young men is thus a process characterized by different kinds of physical and emotional struggles. They are confronted with hardship both in their social and affective lives.

Recyclers work long hours. They get up in the early hours of the day to roam the city for the precious waste they trade (Inverardi-Ferri 2018b). They are not only subjected to the stigma of being migrants, but also to the shame of making a living from the refuse of other people. They are at the bottom of the urban social structure. Yet for many the ability of enduring such a life is a source of pride that distinguishes them from the urban population. As Lao Huo, a young migrant in his early 20s put it in our personal communication, their work is tough, but it is worth doing, because it provides many services to the local community. Migrants recycle waste, they create job opportunities for fellow villagers, and they provide cheap secondhand goods, such as refurbished refrigerators and washing machines, for people in the city. In the words of Lao Wei: "We don't steal and we don't snatch. It's an honest job."

Over the years, the growing number of migrants engaging in the waste industry resulted in the production of a specific economic geography. Already towards the end of the 1990s waste pickers in Beijing were so numerous that several agglomerations emerged at the urban-rural fringe of the metropolis in the form of informal settlements (Béja et al. 1999; Tang and Feng 2000). As suggested in a personal communication with a local businessman, in the 1980s and 1990s the scale of these settlements was relatively small, with a few warehouses that gathered recycling workshops under the same roof. Recyclers were literally living and carrying out their activities side by side in a few warehouses; however, in a couple of decades, the industry, embedded into family and kinship structures, grew at a fast pace. Having gone back to the home village to visit their families, recyclers returned to Beijing with younger relatives and friends. In this way, nourished by new generations of migrants, the industry quickly developed into a complex system, today composed of many buyers, suppliers, and facilitators. This includes individual collectors, manufacturing companies, secondhand shops, formal recycling companies, producers, complementary businesses, logistics specialists, and other actors operating both locally and nationally (Inverardi-Ferri, 2017a). Roaming in urban neighborhoods, an army of hundreds of thousands of individual collectors—equipped with little except for tricycles or scooters—collect recyclables that are then resold to local "street trading points," which in turn trade with middlemen in suburban "waste markets" (Linzner and Salhofer 2014). The terms 'waste market' and 'recycling markets' are adopted to define large agglomerations of enterprises trading and processing recyclables, usually situated at the rural-urban fringe of metropolises in China.

As agglomeration economies, these markets provide many opportunities for young people who can take advantage of the social networking of the migrant community that inhabits these places. This was the case for Xiejia, an agglomeration specializing in the refurbishment of used electronics. Established in the northern suburbs of Beijing, the Xiejia settlement was demolished in 2015. Among my informants were two brothers, in their early 20s. At the time of my research, one of the brothers provided a number of logistic and administrative services to the community of recyclers established in the Xiejia area. His brother had a small secondhand shop where refurbished electronics were sold to the public. Although they specialized in different segments of this business, for both young men the market of Xiejia provided the means for their passage to adulthood. The demolition of these informal settlements, often due to urban redevelopment projects, represents a significant threat for young migrants engaging in waste activities, as I show in the following paragraphs with my analysis of Dongxiaokou.

Dongxiaokou, a former rural village in the northern periphery of Beijing not far from the fifth ring road (around 10 kilometers away from the city center), developed in a few years into the biggest waste market of the metropolis, with thousands of family-owned enterprises established in the area (Inverardi-Ferri 2018a; Tong and Tao 2016). Here every kind of discarded item could be traded daily; however, as was the case with Xiejia, the market of Dongxiaokou was demolished at the beginning of 2015 due to an urban redevelopment project aimed at reorganizing the neighborhood into a residential area (Inverardi-Ferri 2018a). This agglomeration produced many employment opportunities. Yet the local administration prioritized the new imperatives of urban development over the established recycling activities performed at this site, and recyclers had to close down their businesses. As the son of a wealthy merchant in Dongxiaokou explained during an interview:

> Though we produce a lot of job opportunities, our government doesn't attach much importance to employment. Blanche[2] said that the jobs created by waste sorting and recycling are several times the number of jobs created by incinerators and landfills. This industry is atrophying, although more and more people are starting to pay attention to it. Everybody knows that [recycling] is a sunrise industry and would like to invest into it. However, the government, in reality, is squeezing our space.

Indeed, the demolition of Dongxiaokou became a new source of insecurity for many young migrants who had found different types of employment there. When early rumors of demolition spread several years ago, the informant quoted above started to look for a site where he could relocate his activity and invest in upgrading his technology. He quickly identified a new site. At the time of this research, however, the project had not yet started. While administrative obstacles had blocked the creation of the new activity, the informant explained that the root of the problem was a political one.

Undoubtedly, the waste industry has produced many job opportunities for young migrants. But recycling activities that were tolerated or even promoted by authorities as a way to boost local development in previous decades nowadays are stigmatized, and often discursively portrayed as illegal practices (Lora-Wain-wright 2016). The space produced by this economic phenomenon is therefore subjected to an appropriation by formal urbanism and exposed to new rounds of enclosure. In line with the new imperatives of "global cities," authorities also promote new modes of waste management based on large-scale and mechanical industry that, not surprisingly, results in a dispossession of the "waste commons" for traditional actors in the sector (Inverardi-Ferri 2018a), reducing job opportunities for new generations of migrant workers.

Waste activities thus appear to be subjected to a continuous tension between the phenomena of informality and unemployment. Unemployed young people find in informal activities a safety net to make a living; however, the precarious conditions of this industry result in constant threats to a person's right to perform recycling activity. These threats come from various sources such as new regulatory frameworks, evictions, or other forms of enclosure. As suggested by Ferguson (2013: 231), in performing these kinds of daily practices young migrants "are left in a very precarious position—hanging (as the literal meaning of 'depend' suggests) by a thread (or perhaps, in the better case, by a frail network of threads)."

A political economy of variegated capitalism

This chapter has sought to contribute to recent geographical research on youth unemployment in the global South through an empirical account of scrap traders in Beijing. It has brought research on youth unemployment into the conversation with scholarship on labor geography. It has mapped, through ethnographic material, the emergence of spaces of waste in contemporary Beijing as a way to develop an appreciation of the different survival strategies utilized by young migrants and an understanding of the consequences for shaping different labor regimes. It concludes with a summary that makes explicit the relevance of investigating unemployment from the perspective of informality in order to analyze the political economy of contemporary capitalism.

Unemployment is a process that not only describes the loss of a job but also a condition that involves the larger dimensions of personhood and social life. The hardship of finding a job in the context of global changes and economic transformations is a phenomenon that nowadays many young people experience during their passage to adulthood. It thus involves both the sphere of production and the most intimate dimensions of social reproduction (Jeffrey and Mcdowell 2004). The choice of daily practices and survival strategies deployed to overcome the condition of unemployment has implications that go beyond a mere understanding of individual experiences to shed light on the kind of social relations contested and constructed through specific actions.

This chapter has suggested that informality, as a way to overcome unemployment, is an integral part of the diverse dimensions of the economic geography of the global South (Jessop 2011; Peck and Theodore 2007; Yeung 2004; Yeung and Lin 2003) and the mainstay in the production of heterogeneous labor regimes (Herod 1997). Modes of fall-back work and getting by are not always a forced choice, but often a conscious strategy to make a better living in conditions that are considered less coercive than formal waged work. Informal practices therefore can be understood as a way to regain control over labor time (Coe and Jordhus-Lier 2011) and resist and rework more oppressive capitalist social relations (Katz 2004).

This chapter has fostered an understanding of informal practices as a strategy for many young women and men to improve their material means and enable agency. The study of informal practices can serve to shed light on how individual choices are embedded into the geographies of specific communities and labor markets. The comparative study of these phenomena therefore can greatly promote our understanding of the connection between youth unemployment and informality in different geographical contexts (Jeffrey 2008). It can serve to show how these phenomena are entangled within wider social relations (Coe and Jordhus-Lier 2011) and help to investigate how individual and collective choices to overcome the condition of unemployment are not only a mechanical outcome of structural phenomena but also a transformative element of the political economy of different geographies.

Notes

1 Names of informants have been changed in the text to ensure their anonymity.
2 An environmental scholar. The name has been changed in the text.

References

Bai, Limin. 2006. "Graduate Unemployment: Dilemmas and Challenges in China's Move to Mass Higher Education." *The China Quarterly*, 185: 128–144. doi: 10.1017/S0305741006000087.

Béja, Jean-Phillipe, Michael Bonnin, Tang Can, and Feng Xiaoshuang. 1999. "Comment Apparaissent les Couches Sociales. La Différenciation Sociale chez les Paysans Immigrés du 'Village du Henan' à Pékin" ["How Social Layers Appear. Social Differentiation among Immigrant Farmers from 'Henan Village' in Beijing"]. *Perspectives Chinoises*, 52: 30–55.

Cheng, Yi'En. 2016. "Critical Geographies of Education Beyond 'Value': Moral Sentiments, Caring, and a Politics for Acting Differently." *Antipode*, 48(4): 919–936. doi: 10.1111/anti.12232.

Coe, Neil. 2013. "Geographies of Production III: Making Space for Labour." *Progress in Human Geography*, 37(2): 271–284. doi: 10.1177/0309132512441318.

Coe, Neil, and David Jordhus-Lier. 2011. "Constrained Agency? Re-evaluating the Geographies of Labour." *Progress in Human Geography*, 35(2): 211–233. doi: 10.1177/0309132510366746.

Coe, Neil, and Henry Wai-chung Yeung. 2015. *Global Production Networks: Theorizing Economic Development in an Interconnected World*. Oxford: Oxford University Press.

Dong, Madelein Yue. 1999. "Juggling Bits: Tianqiao as Republican Beijing's Recycling Center." *Modern China*, 25(3): 303–342.

Dong, Madelein Yue. 2003. *Republican Beijing: The City and Its Histories*. Berkeley, CA: University of California Press.

Ensmenger, Devone, Joshua Goldstein, and Richard Mack. 2005. "Talking Trash: An Examination of Recycling and Solid Waste Management Policies, Economies, and Practices in Beijing." *East-West Connections*, 5(1): 115–133.

Evans, Bethan. 2008. "Geographies of Youth/Young People." *Geography Compass*, 2(5): 1659–1680. doi: 10.1111/j.1749-8198.2008.00147.x.

Fan, C. Cindy 2008. *China on the Move: Migration, the State, and the Household*. Abingdon, UK: Routledge.

Ferguson, James. 2013. "Declarations of Dependence: Labour, Personhood, and Welfare in Southern Africa." *Journal of the Royal Anthropological Institute*, 19(2): 223–242. doi: 10.1111/1467-9655.12023.

Gold, Thomas, William Hurst, Jaeyon Won, and Li Qiang. 2009. *Laid-Off Workers in a Workers' State: Unemployment with Chinese Characteristics*. Basingstoke: Palgrave Macmillan.

Goldstein, Joshua. 2006. "The Remains of the Everyday: One Hundred Years of Recycling in Beijing." In *Everyday Modernity in China*. M.Y. Dong and J. Goldstein, eds. Seattle, WA: University of Washington Press, pp. 260–302.

Harvey, David. 2006. *Spaces of Global Capitalism: Towards a Theory of Uneven Geographical Development*. London: Verso.

Herod, Andrew. 1997. "From a Geography of Labor to a Labor Geography: Labor's Spatial Fix and the Geography of Capitalism." *Antipode*, 29(1): 1–31. doi: 10.1111/1467-8330.00033.

Hudson, Ray. 2016. "Rising Powers and the Drivers of Uneven Global Development." *Area Development and Policy*, 1(3): 279–294. doi: 10.1080/23792949.2016.1227271.

Hurst, William. 2009. *The Chinese Worker after Socialism*. Cambridge: Cambridge University Press.

Inverardi-Ferri, Carlo. 2017a. "Commons and the Right to the City in Contemporary China." *Made in China*, 2: 38–41.

Inverardi-Ferri, Carlo. 2017b. "Variegated Geographies of Electronic Waste: Policy Mobility, Heterogeneity and Neoliberalism." *Area Development and Policy*, 2(3): 314–331. doi: 10.1080/23792949.2017.1307091.

Inverardi-Ferri, Carlo. 2018a. "The Enclosure of 'Waste Land': Rethinking Informality and Dispossession." *Transactions of the Institute of British Geographers*, 43(2): 230–244. doi: 10.1111/tran.12217.

Inverardi-Ferri, Carlo. 2018b. "Urban Nomadism: Everyday Mobilities of Waste Recyclers in Beijing." *Mobilities*. doi: 10.1080/17450101.2018.1504665.

Jeffrey, Craig. 2008. "'Generation Nowhere': Rethinking Youth through the Lens of Unemployed Young Men." *Progress in Human Geography*, 32(6): 739–758. doi: 10.1177/0309132507088119.

Jeffrey, Craig. 2010. "Geographies of Children and Youth I: Eroding Maps of Life." *Progress in Human Geography*, 34(4): 496–505. doi: 10.1177/0309132509348533.

Jeffrey, Craig. 2012. "Geographies of Children and Youth II: Global Youth Agency." *Progress in Human Geography*, 36(2): 245–253. doi: 10.1177/0309132510393316.

Jeffrey, Craig. 2013. "Geographies of Children and Youth III: Alchemists of the Revolution?" *Progress in Human Geography*, 37(1): 145–152. doi: 10.1177/0309132511434902.

Jeffrey, Craig, and Jane Dyson. 2013. "Zigzag Capitalism: Youth Entrepreneurship in the Contemporary Global South." *Geoforum*, 49: R1–R3.

Jeffrey, Craig, and Linda Mcdowell. 2004. "Youth in a Comparative Perspective: Global Change, Local Lives." *Youth and Society*, 36(2): 131–142. doi: 10.1177/0044118x04268375.

Jessop, Bob. 2011. "Rethinking the Diversity of Capitalism: Varieties of Capitalism, Variegated Capitalism, and the World Market." In *Capitalist Diversity and Diversity within Capitalism*. G. Wood and C. Lane, eds. London: Routledge, pp. 209–237.

Katz, Cindi. 2004. *Growing Up Global: Economic Restructuring and Children's Everyday Lives*. Minneapolis, MN: University of Minnesota Press.

Kirby, Peter, and Anna Lora-Wainwright. 2015. "Exporting Harm, Scavenging Value: Transnational Circuits of E-waste between Japan, China and Beyond." *Area*, 47(1): 40–47. doi: 10.1111/area.12169.

Lee, Grace, and Malcolm Warner. 2007. *Unemployment in China: Economy, Human Resources and Labour Markets*. London: Routledge.

Lefebvre, Henri. 1991. *The Production of Space*. Oxford: Blackwell. First published 1974.

Linzner, Roland, and Stefan Salhofer. 2014. "Municipal Solid Waste Recycling and the Significance of the Informal Sector in Urban China." *Waste Management and Research*, 32: 896–907. doi: 10.1177/0734242x14543555.

Lora-Wainwright, Anna. 2016. "The Trouble of Connection: E-waste in China between State Regulation, Development Regimes and Global Capitalism." In *The Anthropology of Disconnection: The Political Ecology of Post-Industrial Regimes*. I. Harper, S. Vaccaro, and S. Murray, eds. New York: Berghahn, pp. 113–131.

Meagher, Kate. 2013. "Unlocking the Informal Economy: A Literature Review on Linkages between Formal and Informal Economies in Developing Countries." Women in Informal Employment: Globalizing and Organizing (WIEGO) Working Paper No. 27. Retrieved from www.wiego.org/sites/default/files/publications/files/Meagher-Informal-Economy-Lit-Review-WIEGO-WP27.pdf

Minter, Adam. 2013. *Junkyard Planet: Travels in the Billion-Dollar Trash Trade*. New York: Bloomsbury Publishing.

Murphy, Rachel, and Ran Tao. 2007. "No Wage and No Land: New Forms of Unemployment in Rural China." In *Unemployment in China: Economy, Human Resources and Labour Markets*. G.O.M. Lee and M. Warner, eds. London: Routledge, pp. 128–148.

Peck, Jamie, and Nik Theodore. 2007. "Variegated Capitalism." *Progress in Human Geography*, 31(6): 731–772. doi: 10.1177/0309132507083505.

Polanyi, Karl. 1944. *The Great Transformation: The Political and Economic Origins of Our Time*. Boston, MA: Beacon Press.

Qian, Junxi. 2015. "No Right to the Street: Motorcycle Taxis, Discourse Production and the Regulation of Unruly Mobility." *Urban Studies*, 52(15): 2922–2947. doi: 10.1177/0042098014539402.

Schucher, Günter. 2014. "A Ticking 'Time Bomb'?—Youth Employment Problems in China." German Institute of Global and Area Studies (GIGA) Working Papers No. 258, GIGA Research Unit: Institute of Asian Studies. Retrieved from www.giga-hamburg.de/en/system/files/publications/wp258_schucher.pdf

Schucher, Günter. 2015. "The Fear of Failure: Youth Employment Problems in China." *International Labour Review*, 156(1): 73–98. doi: 10.1111/j.1564-913X.2015.00048.x.

Skelton, Tracey. 2009. "Children's Geographies/Geographies of Children: Play, Work, Mobilities and Migration." *Geography Compass*, 3(4): 1430–1448. doi: 10.1111/j.1749-8198.2009.00240.x.

Smith, Chris, and Jenny Chan. 2015. "Working for Two Bosses: Student Interns as Constrained Labour in China." *Human Relations*, 68(2): 305–326. doi: 10.1177/0018726714557013.

Tang, C., and X. Feng. 2000. "河南村' 流动农民的分化" ["The Stratification of Rural Migrants in the 'Henan Village'"]. *Sociological Studies*, 4: 72–85.

Tong, Xin, and Dongyan Tao. 2016. "The Rise and Fall of a 'Waste City' in the Construction of an 'Urban Circular Economic System': The Changing Landscape of Waste in Beijing." *Resources, Conservation and Recycling*, 107: 10–17.

Webber, Michael, and Ying Zhu. 2007. "Primitive Accumulation, Transition and Unemployment in China." In *Unemployment in China: Economy, Human Resources and Labour Markets*. G.O.M. Lee and M. Warner, eds. London: Routledge, pp. 17–35.

Wójcik, Dariusz, and James Camilleri. 2015. "Capitalist Tools in Socialist Hands? China Mobile in Global Financial Networks." *Transactions of the Institute of British Geographers*, 40(4): 464–478. doi: 10.1111/tran.12089.

Yeung, Henry Wai-chung. 2004. *Chinese Capitalism in a Global Era: Towards a Hybrid Capitalism*. London: Routledge.

Yeung, Henry Wai-chung, and George C.S. Lin. 2003. "Theorizing Economic Geographies of Asia." *Economic Geography*, 79(2): 107–128. doi: 10.1111/j.1944-8287.2003.tb00204.x.

Yeung, Henry Wai-chung, and Kris Olds. 2000. *The Globalization of Chinese Business Firms*. Basingstoke: Palgrave Macmillan.

Yu, Hua. 2009. *Brothers*. New York: Pantheon Books.

Zhang, Zhang. 2000. "Mediating Time: The 'Rice Bowl of Youth' in Fin de Siecle Urban China." *Public Culture*, 12(1): 93–113.

11 The youth wage subsidy in South Africa

A controversial proposal to respond to mass youth unemployment

Crispen Chinguno

South Africa faces a problem that intimately links unemployment, inequality, and poverty. The National Planning Commission (NPC)[1] sees this "triple" threat as the most critical challenge facing the country in the post-apartheid context. This chapter addresses one component of that complex challenge facing the country—youth unemployment—with an interesting policy proposal.

On January 1, 2014, the South African government led by the African National Congress (ANC) adopted a wage subsidy policy that would offer economic assistance to employers who hired young workers. The rationale behind this proposal was to boost youth employment by helping to defer the costs of training, education, and skills development that many employers are unwilling to take on. After more than three years of implementation, the results have been ambiguous, with major players debating the pros and cons of the policy. Below I explore the history and implementation of the policy as it exists, and offer some suggestions as to how it may be strengthened.

The unemployment rate in South Africa has remained persistently high, at over 25% in the last three years using the narrow definition and over 30% using the expanded definition.[2] It peaked at 27.7% in the first quarter of 2017; the highest level since 2003 (narrow definition) (Statistics South Africa 2017). One direct outcome of unemployment is the escalation of social tensions and crime. For example, we have seen an escalation of violent xenophobic attacks in many parts of the country in the last decade, some of which have been partly linked to competition for scarce resources including job opportunities (Neocosmos 2010; Landau 2012). The intersection of these problems presents some major post-apartheid socioeconomic and political challenges.

The problem is especially critical when it comes to the youth population. The World Economic Forum "Global Risks" report of 2014 profiled South Africa as having the third highest level (after Greece and Spain) of youth unemployment in the world, at 53.4%. This propelled the National Treasury to propose a youth wage subsidy in 2011 as a direct intervention in the labor market. This radical proposal was not easily accepted by all parties involved, and was only adopted in 2014 after a protracted negotiation process had been initiated. The advocates of the youth wage subsidy argued that it would benefit a specific targeted group within the population; however, the main trade union federation—the Congress of South African Trade Unions (COSATU)—opposed the idea for reasons explored below. This subsidy has been an unprecedented active labor market policy of the South African government. The rationale, as argued by its proponents, is premised on the experience of other countries that have adopted a similar policy. Yet there is much debate as to how to evaluate the data. Do they show that the policy has been detrimental or helped?

As part of the negotiation process, the youth wage subsidy proposal had to be dealt with first by the National Economic Development and Labour Council (NEDLAC), a tripartite body that emerged from the post-apartheid order, with representation from government, organized business, trade unions, and the community; it was designed to reach consensus on issues related to the conception of national social and economic policy. One of its principle objectives is to promote collective and inclusive decision-making processes promoting economic growth and social equity. As was expected, the youth wage subsidy proposal generated tensions within NEDLAC that manifested along ideological lines. COSATU, the largest labor federation in the country and one of the voices of workers at NEDLAC, vehemently opposed the implementation of the subsidy. Conversely, the other side included the South African National Treasury and the opposition party, the Democratic Alliance (DA), all perceived by their opponents as advocates of big capital interests.

I begin now with an overview of how the youth unemployment crisis manifests in South Africa. Then I address the debates about the proposed legislation, and then consider how it could be more effective.

The youth unemployment crisis in South Africa

The South African population, as in many developing countries, is disproportionately young, with more than 40% of citizens in the 15–35-year-old cohort. About one-quarter of the population is below the age of 25. The majority of this population is poorly educated, partly because of the legacy of apartheid, which relegated blacks to poor quality education and into the periphery of the economy. This makes it difficult for many to acquire the skills demanded by the economy and raises questions as well about unemployment, which in the South African context does not occur evenly across all the age cohorts. It is generally more acute within the youth age cohort. Youth unemployment (16–24-year-old cohort) in South Africa was 60% in 2012 using the expanded definition of unemployment. It has remained at over 50% for most of the period since 2012 and peaked at 65.7% in 2017. One-third of the young people in South Africa are neither employed nor participating in education or training.

As highlighted by the World Economic Forum (2014), South Africa has one of the highest levels of youth unemployment in the world. Recent experience drawn from the Arab uprisings has shown how the problem of youth unemployment can gradually transmute into a major political crisis followed by reform. In many countries, young people generally have a history of having a very high level of political engagement and are often the protagonists of transformation in many societies (Chinguno et al. 2017). Their participation in politics is usually driven by their high aspirations for a new order. As a result, this participation attracts special attention from many governments.

The South African economy maintained positive growth rates throughout the period from 1994 to 2013; however, this growth in the economy was not matched by proportionate job growth. As a result, many young people face problems finding employment, being new to the job market and lacking experience. Although youth unemployment was a problem before the post-1994 dispensation, there was no special focus on it. It only became an issue after the 1994 democratic transition and even more acute following the 2008 global recession. It remained persistently high well after this global crisis.

Using the International Labour Organization (ILO) measure of the youth category (those aged 15–24), there are more than 1.2 million unemployed young workers (30% of the overall unemployment) in South Africa. This translates to a youth unemployment rate of 52% or, stated another way: one in every two people under the age of 25 looking for work is jobless. This is exacerbated by the fact that South Africa has more people on social grants than in employment. In 2016 there were 17 million people on social grants compared to 15.5 million in employment (South African Institute of Race Relations [SIRR] 2016). COSATU and other civic organizations have often cited the problem of youth unemployment as being a ticking time bomb that demands urgent attention.

Part of the reason why unemployment is so high in the youth age cohort is structural and endemic. The South African education system, for example, produces poorly skilled graduates who cannot readily fit into the labor market. There are many cases in which employers have complained of a skills gap and argued that a lot of the graduates are unemployable (National Treasury 2011; ILO 2013). In addition, many young people exit the education system prematurely (before high school), making it very difficult for them to be employed or access vocational or other forms of tertiary training—60% of those who are unemployed in South Africa have no secondary education. A shocking 95% of unemployed young people have no tertiary education. This clearly reflects a relationship between youth unemployment and the problem of poor education and training. Higher-level education and training enhance the chances of being employed for all age cohorts and genders (Benya et al. 2017).

Paradoxically, South Africa faces a problem of youth unemployment at the same time as it experiences a crisis-level supply of skilled labor. There is a gross mismatch between the type of jobs on the market and the skills readily available to fill them. For example, South Africa has been facing a critical shortage of craftspeople in the past decade. The average age of a craftsperson is about 55; most of them are white and nearing retirement. Currently, there is a shortage of between 30,000 and 40,000 craftspeople in the South African economy, a situation tied to the poor education system (Benya et al. 2017). For instance, the system produces very few graduates with an aptitude for mathematics and science, two requisite subjects in the training of craftspeople.

Although six million South Africans are unemployed, more than 800,000 highly skilled positions remain unfilled because of a shortage of workers with the training needed to fill them (Benya et al. 2017). This shortage is an obstacle to economic growth and negatively impacts job creation and expansion of businesses. And yet most of the unemployed (at least 60%) are blacks with no secondary education in mathematics and science, and thus they may not be easily trained to close the skills gap. Most employers generally perceive young people as expensive to employ because they lack the requisite experience and because there is a gap between their level of productivity and the cost of employing them.

The other reason for acute youth unemployment is tied to the lethargic performance of the economy in recent years. As a result of both exogenous and endogenous factors, the South African economy has been in phases of weak economic growth, but, for the first time since the democratic transition, it plunged into a technical recession in the first half of 2017. The economic growth rate in most of the 23 years after the historic 1994 democratic transition has been below the target set by the market-oriented economic blueprint of the Growth, Employment and Redistribution (GEAR) policy as projected by the South African Reserve Bank. This subdued economic growth partly accounts for the general scarcity of job opportunities.

Many of the young people in South Africa are part of the discouraged workforce, those in the economically active age groups who have given up looking for work. It costs money to look for work, and many young people, convinced that there are no jobs, thus stop looking. In some cases the jobs may be available but at a wage below the cost of what makes going to work economically feasible. For example, as a result of apartheid's spatial geography, most black people in Johannesburg live in townships far from the economic hub of the city where jobs are located, and transportation is prohibitively expensive and arduous. This geography defines most urban settlements in South Africa.

In past years the South African economy has structurally been shifting from primary production—agriculture and mining—to a service economy. This shift has not only reduced the job growth rate but has disadvantaged workers who are less skilled and have a low literacy level. The overall number of job opportunities in the economy has declined significantly because of this structural shift. While the structural change in the economy is expected and ongoing, the problem in the South African context is that many workers have failed to adjust to the new demands, and their skills, education, and experience have become obsolete. The new world of work inevitably demands workers who can continuously adopt (and adapt to) new skills.

As a result of market demands and the pressure to maximize productivity, most employers have shifted to flexible forms of employment. Most of these new jobs are precarious and short term, and usually are taken up only by the young. Many young people move in cycles between employment and unemployment because of the precarious nature of available jobs. Moreover, they are restricted to the entry-level jobs that are sensitive to seasonal fluctuations. Youth unemployment is thus very sensitive to structural changes and business cycles. If there is a crisis most companies will stop hiring or let the least skilled and experienced workers go. In many cases this affects young people more than any other age cohort—young people, the ones usually entering the job market for the first time, are less employable compared to older adults because they do not have the adequate job experience in many cases.

The history of South Africa is embedded in the politics of very strong and politically active trade unions. The nation had the fastest unionization rates of anywhere in the world in the 1990s, when the labor movement globally was in a serious state of decline. Trade unionization in South Africa peaked at 40% in the 1990s but declined to 29% by 2013 (COSATU 2013). In comparison with other countries, trade unions in South Africa are still relatively strong both politically and economically and have influence extending to the broader society. This historical context has convinced many liberals that the problem of unemployment in South Africa is partly a result of the power and political influence of the trade unions. According to this theory, wages in most sectors in South Africa have reached very high levels because pressure from strong trade unions has made it expensive to hire labor in general. As a result, many employers restrain from hiring and scale up the level of their technology, reducing the amount of necessary human labor required. Young people are thus most likely to be employed by companies that are not unionized and in jobs that are short term and precarious. The 1994 democratic transition enhanced the political leverage of trade unions in South Africa given their role during the struggle for democratization. COSATU is in a formal alliance with the ruling political party, the ANC. The post-apartheid period has seen a general increase in real wages and job security for the sectors with a high trade union density. This high-level political strength of trade unions has been cited as one of the reasons why employers are reluctant to hire new workers.

There is a general perception that the post-apartheid labor market, characterized by difficulties in hiring and firing, is too rigid. Other factors also propel employers to adopt capital-intensive production methods that reduce the labor head count. This in many ways reduces the number of job opportunities, especially those that require elementary skills.

The problem of youth unemployment in South Africa is complicated by the politics of difference: gender, race, and ethnic dimensions. Youth unemployment is far more a problem for blacks and women than it is for whites because of the structural problems tied to the legacy of apartheid. Drawing on the gender dimension, youth unemployment for young women was at 57.3% compared to 48.3% for young men in 2017 (Statistics South Africa 2017). Figure 11.1 shows how between 2008 and 2012 the youth unemployment rate of blacks using the expanded definition remained high at over 50% throughout that period when compared to that of whites, which was below 20% for the same period. The Indian/Asian and colored levels of youth unemployment for the same period were between these two extremes (see Figure 11.1).

Two-thirds of black young people in South Africa are unemployed compared to one in 50 whites in the same cohort. The problem is even more acute for black

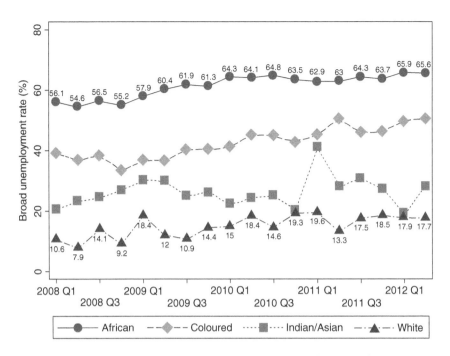

Figure 11.1 Race and youth unemployment in South Africa (Levinsohn et al. 2014)

women who face an intersection of challenges: race, gender, culture, and class simultaneously. Black young people face challenges in getting employment and conversely may easily lose their jobs compared to other races. The problem is structural, and tied to the apartheid legacy and the hetero-patriarchal order that characterize South African society. It is therefore imperative to employ intersectional lenses when analyzing the problem of youth unemployment in South Africa.

The mechanics of the wage subsidy

As outlined above, young people occupy a precarious position in the labor market. They are usually the last choice when job opportunities arise and first out when there is retrenchment (ILO 2013). Senior employees are usually more expensive to get rid of because of the retrenchment regulations and/or collective bargaining agreements that put a premium price on their termination of service by the employer. There are other reasons why young people may not be the most desirable for many employers. Employers prefer workers with experience and skills and often are not so keen to cover the cost of education and training. Moreover, in South Africa employers usually scout for high-level skills, yet the education and training system is poor and may not reflect the productivity potential of a worker. The poor education system feeds into poor workplace learning facilities. This is one of the reasons why South Africa currently faces a critical shortage of craftspeople. The potential of new entrants is clouded with uncertainty because of the poor education system and, as a result, most employers consider new workers to be a high risk for hiring. This puts a high premium on those with experience and skills.

Literature on international experience with unemployment strongly suggests that the problem of youth unemployment can be addressed by a direct and vigorous labor market policy intervention. One strategy is to reduce the cost of employing young people to make their employment more attractive. This applies to training costs associated with recruiting young and inexperienced workers and takes cognizance that there is a gap between productivity and the wage rate for young people. It also addresses the neoclassical economists' argument on the mismatch between the youth entry level of productivity and the one deemed optimum. The wage subsidy thus reduces the cost of hiring inexperienced young people. Drawing from a neoclassical perspective, the National Treasury (2011) argued that a gap between real wages and productivity undermined competitiveness and discouraged businesses from hiring workers, and this in turn raised the rate of unemployment in South Africa. The National Treasury (2011) thus proposed and justified the introduction of a youth wage subsidy on the basis that this was to narrow the gap between entry-level real wages and productivity for young people.

Policies that support accelerated and sustained economic growth are important because a growing economy boosts labor demand and decent employment opportunities. South Africa created about two million jobs between 2003 and 2008 with an average GDP growth of about 4.9% for the same period. Much of this job creation was concentrated in sectors that enjoyed rapid growth, such as construction (13.9% or 500,000 jobs) and finance (9.6% or 520,000 jobs). The exponential rise in youth unemployment in South Africa initiated debate on its possible cause and what could be done about it. Contributing to this debate in 2011, the National Treasury argued the importance of recognizing that, in an environment where young people have little work experience and the costs of firing and hiring new staff can be high, firms will tend to hire fewer young people than they should. Thus the demand for young people to work in firms is low. First, the National Treasury (2011) proposed a youth employment subsidy as a measure to resolve this problem. It argued that the subsidy would reduce the financial costs or risk associated with not knowing the youth pre-employment productivity level. Second, the National Treasury (2011) argued that the youth employment subsidy could help to make the training of young workers more affordable for employers, particularly for smaller employers. Furthermore, the National Treasury (2011) argued that the subsidy was intended to encourage more active job searches because most young people believe that they will be able to find work.

The basis of this proposal is drawn from the experience of more advanced economies. The then Minister of Finance Pravin Gordhan announced a US$400 million youth wage subsidy in his 2011 budget speech. The idea to introduce this subsidy was conceived by the National Treasury drawing from a proposal by James Levinsohn (2007), who argued that a subsidy would address youth unemployment by lowering the costs associated with matching firms to new workers. He noted that structural unemployment cannot be reduced by macro-economic management or temporary swings in aggregate demand, but must be addressed by policy interventions affecting labor demand or supply; for example, a wage subsidy, a search subsidy, reduced regulations for first jobs, and direct employment by the government (Levinsohn 2007). The international experience informing this proposal is particularly drawn from the United States, the United Kingdom, and Turkey, among other countries. This model was adopted in these countries to motivate employers to hire young people following a reduction in the cost of employing them. In Argentina, for example, a youth wage subsidy coupled with training produced positive results (Levinsohn et al. 2014).

In South Africa the subsidy was introduced as an Employment Tax Incentive (ETI). The scheme proposed that employers paying the full minimum wage to new hires aged 18 to 29 earning less than R6,000 (US$500) per month should be able to reclaim 50% of the cost in the first year and 20% in the second year as a tax rebate effective from October 2014. The objective is to accelerate job creation and to raise employment by reducing the cost of employment and encouraging youth employment. The subsidy was to be paid to successful applicants (employers) for two years with a maximum value of R12,000.

Initially, the trial phase of the subsidy proposal was to run for a period of three years with detailed monitoring and reporting on a quarterly basis. Its continuation was to be subject to an evaluation after three years. The National Treasury estimated that the youth employment subsidy would provide 423,000 new jobs for young and less-skilled people aged 18–29, and it was expected to cost R5 billion in tax expenditure over three years. Net new job creation was estimated to be 178,000 jobs at a cost per job of R28,000. The 18–29 age cohort was targeted because that group has the highest level of unemployment and the majority have never worked at all.

The proposal by the National Treasury was underpinned by the expectation that the subsidy would lower the relative cost of hiring young people and therefore increase the demand for young workers. Moreover, it was expected that work experience and training gained during the period of the subsidy would improve longer-term employment prospects and help young people to ensure a positive work trajectory in the future.

The youth wage subsidy gained the support of the National Planning Commis-sion and was incorporated into the National Development Plan (NDP). The NDP identified other policy interventions to address the problem of unemployment, including skills development programs and the reform of the educational system. The NDP further suggested the potential that unemployment may be managed by reducing youth labor market participation through prolonging the number of years in school for young people. This in turn enhances their opportunities for vocational and other forms of training.

The South African government has also adopted other initiatives to deal with the problem of youth unemployment. It has implemented the Sector Education and Training Authorities (SETAs) designed to provide training needs for specific sectors of the economy. In this plan, the employer receives a tax rebate for workers trained through a learnership or apprenticeship program. Other initiatives include the National Youth Development Agency, which runs placement programs, provides skills training (including training in life skills), and supports entrepreneurs through loans and training. The Department of Labour also provides labor centers throughout the country that help with job searches and career development. In addition, the Expanded Public Works Programme (EPWP) provides short-term job opportunities, often requiring low skill levels, on government projects, whereas the Community Works Programme (CWP) has a broader focus to empower communities through a more holistic approach to job creation. These two schemes aim to promote public employment but do not guarantee the right to work.

The youth wage subsidy has been adopted by many countries as a direct labor market intervention policy and way of limiting the cost of employing young people. A review of the wage subsidy in Switzerland showed a positive correlation of job subsidy with job creation (Gerfin et al. 2005). A similar study in Turkey concluded that wage subsidies increase job opportunities (Betcherman et al. 2010). But these studies also highlight a significant number of deadweight jobs that were created but then were terminated at the end of the subsidy. Groh et al. (2012) reviewed the impact of wage subsidies on female graduates in Jordan and concluded that they had a positive impact on job creation in the short term but this declined once the subsidy period expired.

It is important to note that youth wage subsidies may not be a silver bullet because they do not address problems such as overly high costs of employment caused by rigidities in the market. Their impact on unemployment very much depends on flexibility regarding production factors and on the structural characteristics of the labor market (Go et al. 2010). As a result, the National Treasury adopted this proposal as a multipronged approach to tackle the problem of youth unemployment considering that subsidies would not address the structural challenges of job creation.

COSATU and other critiques

COSATU emerged as the main opponent of the youth wage subsidy proposal submitted by the National Treasury. Drawing from research on the experiences of other countries, COSATU argued that the youth wage subsidy was not the ideal policy to resolve the problem that faced South Africa. According to COSATU, international experience and extant literature revealed that wage subsidies are ambiguous in resolving the triple problem of unemployment, inequality, and poverty. COSATU dismissed the claim that there is a gap between the levels of production and real wages; an argument advanced by the National Treasury. It argued that this might be based on an error given that the National Treasury failed to compute this correlation to substantiate its claim. COSATU argued, drawing from a Marxist perspective, that the problem of young people in South Africa was not unemployment but exploitation. It thus proposed that the government should tax employers to pay off this gap directly to the workers. It basically rejected the basis of the argument made by the National Treasury on the grounds that it lacked empirical evidence, especially drawing from the context of developing countries.

Moreover, the National Treasury, according to COSATU, assumed labor to be the major constraint on job creation. COSATU argued that many young workers are in unorganized sectors and do not enjoy the minimum wage, which is not legally binding. COSATU feared the wage subsidy would create a multi-tier labor market that would promote discrimination and result in the substitution of older workers with young workers. Furthermore, COSATU dismissed the evaluation process of the subsidy. It argued that it was very difficult to measure which jobs would have been created as a result of the subsidy. Otherwise it cautioned that the subsidy would cushion employers with jobs that could have been created anywhere.

COSATU maintained its critical position on the youth wage subsidy more than two years after it was introduced, even after the three-year review of the program. It concluded that the youth subsidy as proposed by the National Treasury was never meant to address the challenges of unemployment and inequality but was geared towards resolving what it called the crisis of profitability for companies.

Experience from other places suggests that the implementation of the youth wage subsidy is usually grounded in a pluralist industrial relations paradigm, being a product of class compromise between capital and labor presided over by the state; however, the way this was negotiated and implemented in South Africa was not a culmination of such class compromise. The state and capital forged ahead and implemented the subsidy without the buy-in from labor. The NEDLAC negotiating process never reached a consensus over this issue. Furthermore, the youth wage subsidy is silent on other social factors such as gender, race, ethnicity, religion, disability, location, and the education and training that are critical in understanding the problem of youth unemployment in the South African context. How these factors intersect is even more important to consider when proposing how the problem might be resolved. The trouble with the proposal by the National Treasury is that its conception and implementation failed to promote inclusion and social cohesion, and thus undermined institutions that underpin social dialogue in the post-apartheid social order.

The policy raised a lot of skepticism and, in particular, questions of whether the subsidy was the ideal way of dealing with these socioeconomic and political challenges culminating from the escalation in youth unemployment. COSATU's main argument was that evidence from international experience was less sanguine on the merits, especially in a developing country context, and proposed that the subsidy be linked to structured workplace training and the expansion of skills development through the extension of further education colleges (FETs) from 400,000 graduates to one million. COSATU argued that this would reduce the youth labor force by extending the number of years in education and training. Furthermore, COSATU argued that the problem of youth unemployment is systemic and tied to the crisis of an education system that is failing to produce the requisite skills demanded by the job market. According to COSATU, 60% of the unemployed in South Africa have no secondary education, and 68% of the unemployed have been unemployed for the past five years or have not worked at all in their lives.[3]

Conclusion

This chapter has highlighted that wage subsidies are part of an active labor market intervention that may be designed in the South African context to address the problem of youth unemployment. There are on the one hand suggestions from international literature and experience that wage subsidies enhance employability and job opportunities for young people. Yet, others conclude that the same literature and experience demonstrate that wage subsidies are less positive and not ideal for resolving the problem of unemployment in some contexts. The contestation over the proposal highlights the paradox of the youth wage subsidy and the ideological contradictions of the principle within the South African ruling party and the state. The National Treasury has been accused of representing the right wing of the state, which its critics perceive to be controlled by white business interests. Ultimately, the outcome of the implementation of the youth wage subsidy in South Africa has been both ambivalent and contested, vindicating claims that it may not be the panacea to the problem of youth unemployment in that context.

The focus on youth unemployment may be viewed as an exaggeration for a particular ideological agenda. Youth unemployment in South Africa is projected as acute in the 15–24-year-old cohort, which also includes new entrants into the labor market, however, many of the young people in this cohort are either in high school, college, or in transition from high school to college. The focus on youth unemployment is thus socially constructed. It may be argued that youth unemployment is an illusion socially constructed to serve a particular political or business agenda, or that no convincing argument exists as to why the unemployment of young people should be viewed as a special problem. The problem with the South African youth wage subsidy is that it pays lip service to the idea that unemployment is not just a youth problem, but rather is a reflection of broader structural challenges.

While youth wage subsidies have been used as a policy option in addressing the problem of unemployment, the empirical evidence on its contribution in the South African context thus far has been ambiguous. International literature and experience suggest that there is no conclusive proof that wage subsidies can improve employment and resolve the socioeconomic and political challenges related to it. One important lesson for South Africa from international experience is that the outcome of the youth wage subsidy is context-specific and must consider other factors such as gender, race, location, and schooling. The contribution of a youth wage subsidy as a policy option in resolving the problem of unemployment, poverty, and inequality is thus clouded in uncertainty but was nevertheless adopted by the South African government. There is no unambiguous proof from other contexts that this indeed would resolve the problem of youth unemployment, as the evidence points to both positive and negative effects in different contexts. Therefore, the question that remains unanswered is whether this policy is ideal in resolving the problem of youth unemployment in the South African context. A preliminary review of the policy shows that in 2008 there were 9.175 million young people and 3.762 million of them were employed. In 2016, after almost three years of the implementation of the youth wage subsidy, there were 10.169 million young people and only 3.357 million of them were employed (Statistics South Africa 2017). The three-year review of this policy suggests that it has so far been indifferent in resolving the problem of youth unemployment; however, a proposal has been advanced to extend the youth wage subsidy by a further three years despite the inconclusive results skewed in favor of employers. Finally, it is important to note that no single policy offers the solution to the problem of unemployment. What is required is a comprehensive set of short-term and long-term policy reforms and initiatives, i.e., a multipronged strategy.

Notes

1 The NPC was established by the South African government in 2010 to develop a long-term vision and strategic plan for the country, known as the National Development Plan.
2 Definitions of narrow and expanded rate of unemployment: Narrow rate of unemployment refers to those unemployed and actively looking for work and the expanded rate includes those unemployed and available to work even if they may not have been actively looking for work.
3 Interview with COSATU Policy Coordinator.

References

Benya, Tame, Crispen Chinguno, and Mario Jacobs. 2017. *Future of Work Initiative Country Report: Work and Society*. Johannesburg: International Labour Organization.
Betcherman, Gordon, N. Meltem Daysal, and Carmen Pagés. 2010. "Do Employment Subsidies Work? Evidence from Regionally Targeted Subsidies in Turkey." *Labour Economics*, 17(4): 710–722.
Chinguno, Crispen, Morwa Kgoroba, Bafana Nicolas Masilela, Boikhutso Maubane, Nhlanhla Moyo et al. 2017. *Rioting and Writing: Diaries of Wits Fallists*. Johannesburg: Society, Work and Development Institute, University of the Witwatersrand.
Congress of South African Trade Unions (COSATU). 2013. "The Youth Wage Subsidy in South Africa: Response of the Congress of South African Trade Unions." Retrieved from www.cosatu.org.za/docs/misc/2013/youthwagesubsidy.pdf (accessed January 23, 2018).

Gerfin, Michael, Michael Lechner, and Heidi Steiger. 2005. "Does Subsidised Temporary Employment Get the Unemployed Back to Work? An Econometric Analysis of Two Different Schemes." *Labour Economics*, 12: 807–835.

Go, Delfin S., Marna Kearney, Vijdan Korman, Sherman Robinson, and Karen Thierfelder. 2010. "Wage Subsidy and Labor Market Flexibility in South Africa." *The Journal of Development Studies*, 46(9): 1481–1502.

Groh, Matthew, Nandini Krishnan, David McKenzie, and Tara Vishwanath. 2012. *Soft Skills or Hard Cash? The Impact of Training and Wage Subsidy Programs on Female Youth Employment in Jordan*. Washington, DC: World Bank.

International Labour Organisation (ILO). 2013. *Global Employment Trends for Youth—2013 Update*. Geneva: ILO.

Landau, Loren B. 2012. *Exorcising the Demons Within: Xenophobia, Violence and State-craft in Contemporary South Africa*. Johannesburg: Wits University Press.

Levinsohn, James. 2007. "Two Policies to Alleviate Unemployment in South Africa." Retrieved from http://citeseerx.ist.psu.edu/viewdoc/download?doi=10.1.1.507.8322&rep=rep1&type=pdf (accessed January 23, 2018).

Levinsohn, James, Neil Rankin, Gareth Roberts, and Volker Schöer. 2014. "Wage Subsidies and Youth Employment in South Africa: Evidence from a Randomized Control Trial." Retrieved from http://siteresources.worldbank.org/INTDEVIMPEVAINI/Resources/3998199-1286435433106/7460013-1357765223620/Gareth_Roberts_Presentation.pdf (accessed January 23, 2018).

National Treasury. 2011. *Confronting Youth Unemployment: Policy Options for South Africa*. Pretoria: National Treasury.

Neocosmos, Michael. 2010. *From "Foreign Natives" to "Native Foreigners": Explaining Xenophobia in Post-Apartheid South Africa*. Oxford: African Books Collective.

South African Institute of Race Relations (SIRR). 2016. "More South Africans Receive Grants than Have Jobs—A Recipe for Chaos and Violence." Retrieved from http://irr.org.za/reports-and-publications/media-releases/more-south-africans-receive-grants-than-have-jobs-2013-a-recipe-for-chaos-and-violence/view (accessed January 23, 2018).

Statistics South Africa. 2017. "Quarterly Labor Force Survey: Quarter 1, 2017." Retrieved from www.statssa.gov.za/publications/P0211/P02111stQuarter2017.pdf (accessed January 23, 2018).

World Economic Forum. 2014. *Global Risks 2014 Report* (9th edn). Cologne: World Economic Forum. Retrieved from http://reports.weforum.org/global-risks-2014 (accessed January 23, 2018).

12 The right to work and the youth unemployment crisis in Spain

Ciro Milione

All constitutions are designed to encapsulate the most urgent hopes and aspirations of a community. The specifically social nature of the Spanish Constitution (SC) depicts a case in which the most basic individual needs of the citizens are satisfied by means of actions taken by public powers. Although the right to work is given profound social consideration in the Spanish legal system, and is enshrined in the SC, the suffering caused by the 2007 financial crisis in Spain has made it clear that no law, whether constitutional or regional, can easily tackle the damaging effects of the economic downturn of the labor market on its own. Spain has witnessed dramatic levels of unemployment over the last eight years, despite numerous legal provisions guaranteeing the right to all aspects of employment. These circumstances have led some authors (e.g., Peces-Barba Martínez 1993) to question the true meaning of the inclusion of the right to work in the SC, given the difficulties experienced by the legal standard when regulating labor in the context of a market economy.

The Labour Force Survey carried out by the National Institute of Statistics has shown a steady rise in unemployment levels over the last few years. While in the fourth quarter of 2007 the unemployment rate nationwide was 8.6%, data for the first quarter of 2013 indicated a rate of 27%. The latest data, however—for the first quarter of 2017—show that unemployment has fallen to 19% (National Institute of Statistics 2017). If we focus on youth unemployment, the figures increase dramatically. The Labour Force Survey for the fourth quarter of 2007 recorded 18.7% unemployment among workers aged under 25, compared with 56.9% in the first quarter of 2013, and 46.5% in the second quarter of 2016 (National Institute of Statistics 2017). According to studies carried out by the European Council, between 2007 and 2013, youth unemployment rates at least doubled in 12 countries of the European Union (EU), Spain being one of them (European Council 2016). It should also be noted that, according to the most recent data published by EUROSTAT, the total youth unemployment rate in the EU-28 stands at 22.3%, while in Spain the rate reached 49.6% (EUROSTAT 2016).

To better understand the alarming nature of this situation, we must compare data from different regions of Spain. A comparative analysis confirms that Andalusia has the highest level of youth unemployment (58.6% for a total of

168,700 people), followed by the Canary Islands with 57.6%, Asturias with 57.1%, and Extremadura with 51.8% (National Institute of Statistics 2017). These high levels of unemployment not only represent constitutional violations, but also a failure to offer the basic ingredients needed for dignity. Lack of access to employment for many young people means that they are unable to build a life for themselves, have no access to housing, and are unable to support themselves financially.

This chapter assesses why youth unemployment is so high in Spain and compares regions within the country, as well as conditions within the wider EU, to learn more about this problem. In particular, it pays special attention to the policy fixes in Andalusia as a possible solution.

Why are there such high levels of youth unemployment in Spain?

Any investigation of the causes of youth unemployment requires an analysis of the socioeconomic situation in Spain from a range of perspectives. Such an analysis, however complex, is essential in order to measure the effectiveness of the solutions proposed by the government. First, the poor performance of the Spanish economy is one major cause of the high levels of youth unemployment. Spain's GDP for 2015 (totaling US$1.198 trillion) is substantially worse than that recorded in 2006 (US$1.265 trillion). This roughly US$72 billion difference suggests a long-term problem of stagnation (Bank of Spain 2018).

An analysis of the economic cycles of recent years shows that a slowdown in the national economy has always coincided with an increase in youth unemployment: this happened in the early 1980s when the rate climbed to 45%, with a similar effect following the economic crisis in the mid-1990s. By contrast, during the period of economic growth that occurred in Spain in 2006, when GDP rose by 4.1%, the rate of youth unemployment fell to 17.8% in the fourth quarter (Bank of Spain 2018). In addition to economic stagnation, other factors have contributed to high youth unemployment rates in Spain: school dropout rates, mismatches between the educational system and the labor market, and ineffectual active labor market policies implemented in earlier years. For example, there is a strong correlation between a higher level of education and a lower risk of being unemployed, as we will see below. Indeed, an unemployed person with a high level of education has a better chance of becoming employed than another with less education. In addition, a higher level of education generally means a better salary, greater stability, and a lower reliance on temporary work (Capsada Munsech 2014). These positive effects for those who decide to invest in their education are especially true for the younger generations. In fact, according to EUROSTAT (2017), young European adults with basic education have a higher unemployment rate, followed by those with upper secondary education and those with higher education.

Different studies, conducted using data for the years 2007 and 2010, show that those who had completed secondary or university education were less likely to remain unemployed: almost 21% less in the case of having a college degree and

15% less in the case of high-school education (Faci Lucía 2011). But young Spanish adults endure unemployment rates double those of their European counterparts regardless of their level of education (García López 2014).

What are the causes?

On top of all this is the related question of the early dropout rate, referring to the percentage of the population (those aged 18–24) with at most lower secondary education and no further education or training. This issue is an especially sensitive one in Spain, as shown by EUROSTAT (2017) statistics, revealing that Spain has higher dropout rates than the rest of the EU-28 countries. In 2011, Spain showed a rate of early leavers from education and training slightly higher than 26% of the population, followed by Portugal with 23% (in 2016, Spain with Malta occupy the first position of this rating, with a 19% early leaver rate).

Although this is a real structural problem in Spanish society, there is evidently a strong connection between increases/decreases in early dropout rates and the economic cycles of expansion in Spain. Recent history has shown this to be true. While in recent years the economic slowdown coincided with a decreasing (but still high) rate of early school dropouts, the acceleration of the Spanish economy in 2004, fueled by the construction industry, prompted a large number of untrained young people to drop out due to the demand for unskilled labor (32.2% in that year [EUROSTAT 2017]). This option, seemingly profitable in the short term, turned out to be disastrous in the end, since, starting in 2011, thousands of those young people have been dragged into unemployment after the real estate bubble burst.

Other factors, such as gender dynamics and family structure, contribute to rising dropout rates too. For example, dropping out of school is a predominantly male phenomenon. Data for the first quarter of 2013 indicate that 27.5% of all male students dropped out, compared with 20.2% of female students. Over the same period, 58.7% of dropouts were male (Serrano et al. 2013). It is also evident that the risk of dropping out increases in families facing financial difficulties. Thus the 2011 *Encuesta sobre Condiciones de Vida* [*Life Conditions Survey*] shows that school dropout rates in Spain rose to 44.5% in those families that reported having "a lot of difficulty" in making ends meet, while barely reaching 7% for families that reported making ends meet with "considerable ease" (Serrano et al. 2013). Beyond gender-related factors, family characteristics affect school dropout rates. For example, over 30% of young Spaniards whose mothers lack education beyond the compulsory minimum level also drop out of school. This figure falls to 4.6% for those with mothers who received higher education (Serrano et al. 2013).

Given these circumstances, Spain has undertaken steps to lower the early dropout rate from the current rate of 20% to 15% by 2020. An example of this is the "Plan para la reducción del abandono educativo temprano" ["Reducing early school dropout plan"], which seeks:

- to promote and guarantee the integral development and success of all citizens in the educational process;
- to implement measures of educational intervention to tackle the school dropout causes; and
- to detect cases of risk and intervene to facilitate the reincorporation of those who have abandoned the educational system prematurely.

Despite having the highest dropout rate in Europe, however, over the last 30 years there has been a cross-generational improvement in educational attainment. This improvement, far from uniform among all age groups of Spanish society, has largely affected university education among the younger age cohorts. On the one hand, Spain is four percentage points above the average in the EU as far as university graduates in the population aged 25–34 years. On the other hand, it ranks first in European rates of young adults aged 20–24 years with out a diploma from primary education or lower secondary education. This is why Spain can be discussed as a paradigmatic example within Europe of educational-level polarization: while numerous workers have a college degree, a large number of individuals have failed to complete the first stage of secondary education. The reasons for this phenomenon are shown in the absence of a model for dual professional training, which would combine the academic curriculum with acquiring skills that would be useful for young people preparing themselves for the labor market. Sure enough, Northern and Eastern European countries with lower rates of youth unemployment have been implementing these models for years, managing to get the business world to actively participate in the training of specialized workers who become highly employable. Inexplicably, these combined efforts by the business world and educational centers are not yet happening in Spain. From 2008 to 2014, less than 20% of Spaniards under the age of 25 combined their studies with a job or professional training, equating to less than half of those in the other EU countries (García López 2014).

The polarization of educational level in times of economic slowdown has two subsequent negative effects: underemployment/over-qualification and a fall in the wage premium associated with education. Spain registers percentages well above the average in Europe in both of these areas. In short, the scarcity of job offers combined with the presence of a considerable number of highly educated young adults in the job market, who cannot find appropriate jobs to fit with their qualifications, make this group (college students, college graduates, and post-graduates) take jobs that would traditionally be done by uneducated workers, with a concomitant fall in wage premium.

As Rahona López (2008: 61) points out: "Within the group of young people aged 16 to 35 years [...] women, foreigners and younger individuals display a greater degree of over-education in their first meaningful employment." In Spain, therefore, almost 40% of young workers say they have had jobs that were below their skill level (EUROFOUND 2012).

As we saw at the outset, the SC of 1978 requires that public authorities play an active role in achieving the goal of full employment. Taking a lead from other places in Europe, many regions of Spain have implemented a number of active labor market policies (ALMPs, which, by definition, consist of measures to create employment) in which young people are especially targeted. Nevertheless, between 2005 and 2009, while spending on ALMPs relative to Spain's GDP surpassed the EU-15 average, ALMP spending per unemployed person (at about €1,740) has been shown to be 12.5% lower than the EU-15 average (Ramón García 2011). This means that ALMPs are not effective in solving the problem of the unemployed, since almost half of these resources are directed at encouraging hiring and employment retention. In contrast, spending on training, integrating, and retraining remains low.

Furthermore, many efforts focus on training people who are receiving unemployment benefits, rather than unemployed people without any subsidy. In 2011, only 34.2% of the costs of training programs for employment were reserved for the unemployed (Ramón García 2011). This strategy is wholly incomprehensible; it is as if political figures have given up on investing in this group. The only possible explanation resides in the higher returns to be had from helping individuals who receive unemployment benefits, who, after finding work, are less of a burden on the state.

Finally, these dramatic and complex economic and social circumstances have fueled the so-called discouraged worker effect. Since the beginning of the economic crisis, the number of young Spaniards who have given up actively searching for work has undergone a significant increase. On a European level, Spain currently has the highest number of "NEETS": young people of both sexes, aged 16–29, who are not in education, employment or training. According to the latest study by EUROFOUND, 26.7% of young Spaniards have literally decided to "throw in the towel," removing themselves from the job market, followed by 26.2% of young Italians and 24% of young Bulgarians (EUROFOUND 2016). Next let us analyze the actions proposed by political figures to deal with such a dramatic social situation.

The response: how Andalusia is trying to tackle the issue

Due to the territorial distribution of power outlined by the SC, the different regional autonomous communities that make up the political scene in Spain are each responsible for promoting economic development in the region they govern. The case of Andalusia is paradigmatic. According to the National Institute of Statistics' (2017) data from the second quarter of 2016, this region has the highest unemployment rate in Spain with 29% (for a total of 1,161,100 people) and the highest youth unemployment rate with 58.6% (for a total of 168,700 people), followed in this last field by the Canary Islands with 57.6%, Asturias with 57.2%, and Extremadura with 51.8%.

During the years of profound economic crisis, the Andalusian regional government has tried to curb the falling levels of youth employment in its territory, thereby fulfilling the mandate from its Statute of Autonomy which establishes the following conditions:

In the exercise of the constitutional right to employment, all persons are guaranteed: a) free access to public employment services; b) access to public employment under conditions of equality, in accordance with the constitutional principles of merit and capability; c) access to occupational training; d) the right to rest and leisure.

(Molina Navarrete 2012: 437)

The youth employment plan (PJE), a program from the Andalusian government, involved an investment of up to €167.5 million to supplement the additional €243 million from the multi-annual EU budget (2014–2020) to fund youth employment projects. Initially, the PJE was implemented through five distinct measures: a monthly payment of €400 a year for young people who had completed vocational and/or university studies; the granting of soft loans, incentives for youth self-employment, and tax subsidies for companies hiring young people for their internationalization departments; and finally, the creation of two websites—one for employment and another for entrepreneurship. Later, the Andalusian Parliament approved the "Programa Emple@Joven" and the "Iniciativa @mprende+." These programs updated the PJE with the implementation of the first phase of the EU's Youth Guarantee system in Andalusia. It ensures that all people aged under 25, whether registered as unemployed or not, receive a specific, quality job offer within four months of completing their education or becoming unemployed. It is worth examining a few aspects of this program.

One part, the "Social and Community Cooperation Plan for the Promotion of Youth Employment," was a measure to encourage Andalusian municipalities to hire young people for the implementation of social projects and community cooperation. It included a total of 1,000 grants to offer young people the chance to complete an internship program in the industry that best suited their professional profile for six months. Further, youth employment bonds promoted access to employment and facilitated integration into the labor market for young graduates. This initiative involved conceding financial support to companies which contracted a person who had been granted the Youth Employment Bond. The duration of the labor contract involved a minimum of 12 months' full time or part time employment. This amounted to €4,800 for every full-time contract, and €2,400 for every part-time contract. Finally, the program included grants for research and development to promote young Andalusian researchers in the labor market by facilitating their transition from university to the private sector, achieved through a nine-month professional internship.

More than a year after "Programa Emple@Joven" and "Iniciativa @mprende+" passed, data on the extent of their execution and results are still not available. But it is worth noting that, since their enactment, youth unemployment rates in Andalusia have been fluctuating between 62% in the first trimester of 2014, at the beginning of the program, and 58% in the last trimester of 2016 (National Institute of Statistics 2017). This figure clearly illustrates the inadequacy of those active labor market policies.

The promises: how new political parties aim to tackle the problem on a national basis

Spanish politics have been going through tumultuous and uncertain times over recent years. The last two elections (held in December 2015 and June 2016) have shown the demise of the two-party system—the People's Party (Partido Popular or PP) versus the Socialist Party (Partido Socialista Obrero Español or PSOE)—that had strengthened the young Spanish democracy over the last 38 years. Today, the national parliament has two new political parties: Unidos Podemos and Ciudadanos. These groups have known how to read—and benefit from—the unease of the people and their frustration with an obsolete representative system, breaking the traditional parliamentary dynamics and challenging the political system founded on alternating between Socialist and People's Party governments.

Today the dialog between political parties, which is so crucial to a smooth-running executive branch, is conspicuously absent in Spain. Currently, the government of the People's Party has become stronger than other more progressive alternatives, due to internal fragmentation within the Socialist Party and to the lack of willingness on the part of the left-wing Unidos Podemos. With such political turmoil, it is easy to predict that Spanish society will once again end up without appropriate and effective solutions to deal with the problem of unemployment, as it lacks a parliament inclined to engage in dialog and work on solutions together. And yet, these same political figures that make up the array of parliamentary representation have each highlighted their concern about the job crisis and proposed solutions—ultimately promises—of all kinds to deal with this social catastrophe.

Unidos Podemos, an anti-capitalist party, was born from the movement of social outrage that started on May 15, 2011. This party includes in its political program proposals directed against tax fraud, guaranteeing public services (health, education, etc.), the creation of new systems of production, and economic aid for families and small- and medium-sized companies. As far as employment is concerned, Unidos Podemos has two relevant proposals. One is to repeal the two previous labor reforms (one by the Socialist Party in 2010 and the other by the People's Party in 2012) to de-incentivize hiring only temporary workers. Second, this party proposes lowering taxes on new freelance workers in line with a system of progressive taxation. As far as the poorer workers are concerned, Unidos Podemos proposes a "complementary income program," so that no employee earns less than €900 a month in job benefits. Likewise, this party makes a motion to reduce the working week to 35 hours and reduce overtime in order to encourage new hires (Andalucía Podemos 2015).

Ciudadanos, on the other hand, proposes introducing only one kind of employment contract, which would be both stable and fixed term, and so would repeal the kinds of precarious and part-time contracts that currently exist. Implementing this measure would mean a wave of dismissals, however, which is why it also proposes creating an insurance system to handle that reality. At the same time, there would be awards and incentives for companies that dismiss less and hire more. In order to shorten situations of long-term unemployment, Ciudadanos proposes training programs that would teach skills required by the job market and a minimum income for all those who take part in said training programs and have no other form of income. It also proposes computer literacy and/or language programs for unemployed adults. Finally, Ciudadanos aims to cancel freelance workers' fees if the amount of profit they earn amounts to less than the minimum wage (Ciudadanos 2015).

The Socialist Party proposes eliminating all the tax reductions aimed at fostering new hires in order to reinvest them in employment training, implementing a series of measures aimed at uncovering cases of unreported employment, raising the minimum wage, creating a new Workers' Statute, and creating a subsidy for unemployed adults aged over 52 years of age. As far as contracts are concerned, it proposes three different kinds: one for permanent employment to cover stable positions, one for temporary employment to cover temporary posts, and one for substitutions and training.

Last, but not least, the electoral program of the People's Party discusses the implementation of a model of labor relations based on flexibility, in which dismissal would be the last resort when facing adverse economic circumstances. For freelance workers, this party proposes new regulations in keeping with the situation of each entrepreneur, with the possibility to postpone tax payments when necessary. Also, it proposes measures to facilitate a better work/life balance such as a longer paternity leave, more telecommuting options, and business incentives. Out of all the electoral programs, the People's Party is the only one that contains plans for specific instruments that would tackle youth unemployment by means of ad hoc training courses designed to foster youth entrepreneurship.

The measures proposed by the four main political parties in Spain are conspicuously vague. These electoral programs do not sufficiently outline what the training courses they propose will consist of, nor do they explain how these differ from the active labor market policies over the last 20 years. Likewise, they do not describe specifically what the different reforms of the Workers' Statute will be, what the measures of flexibility will be for the new models of labor relations or for freelance workers, or the ways in which the minimum or complementary income will be financed, which have been emphatically stressed. Certainly the weakness of the current government, the schisms in parliament, and the lack of willingness to commit to problem-solving, make it unlikely that any of these groundbreaking proposals listed above could actually come into effect.

Conclusion

The Spanish state is constitutionally required to play an active role in achieving a number of objectives that are difficult to fulfill, among them, the eradication of inequality and maximization of common welfare. The state assumes the role of a mediator between economic agents, facilitating the negotiation of solutions jointly agreed upon by employers and employees. Similarly, in a system of mass production, the state plays a key role in stimulating demand and in stabilizing employment, wage relations, and so forth (Contreras Peláez 1994). The current situation falls somewhat short of these ideal scenarios. The slowdown in the European economy has produced notoriously devastating consequences throughout the European continent, which have been particularly serious in countries such as Greece, Portugal, Ireland, and Spain, among others (EU 2016).

Set against this background, a number of theories have flourished, claiming that there has been an inexorable "crisis of the welfare state." These claims are based on the delays and inefficiency that affect the provision of basic services, on job insecurity, and on the difficult prospects that lie in store for younger generations (Agudo Zamora 2007). But the attribution of the failings of welfare states to an alleged inability does not constitute a well-founded argument. Let us consider recent events, for example, the situation facing the Greek welfare state, with the evident will of the Eurozone members to prioritize economic logic over the most basic social requirements of the Greek people. This case does not appear to exemplify the hierarchy of values of the EU political system. So, beyond the inability to fulfill the objectives of the welfare state, which may be due to current economic factors, it seems evident that a clear political desire is taking shape in Europe that is aimed at changing the social model. Unfortunately, recent politics have shown that, in certain national contexts, this intense wish for change has mobilized political powers more inclined to populism and demagoguery than to searching for real and effective solutions. The result of the referendum on whether the United Kingdom is to remain in the EU (Brexit) and the relative (albeit unexpected) success of the UK Independence Party can be explained, to a large extent, by the British people's frustration and distrust of their national and European governments, which made a deceptive message seem attractive. Notwithstanding the fact that Europe was the cradle of social democracy from the second half of the 20th century onwards, and notwithstanding the fact that this ideal of coexistence has served to consolidate the concept of citizenship by virtue of defending social rights, the current political agenda still seems to point towards a different set of priorities, with a slow but inexorable change in the paradigm.

This new socioeconomic model involves the implementation of measures inspired by the most hardline forms of neoconservatism and neoliberalism. Thus, while a categorical reduction in public spending and the burden of taxation is welcomed as a means of securing a desired increase in levels of competitiveness, there has also been a dramatic reduction in the social roles of the European States. As Fernández García and Andrés Cabello point out, this economic crisis "has been a great opportunity to take steps that threaten social policies, with certain measures that have acted against social cohesion and solidarity, failing to address society's demands and needs" (2015: 122). All this can only take place when the European institutions clearly understand that any proposal of an economic character (a reduction in public spending, controlling public debt, improving competitiveness, etc.) is subordinate to a goal that is unquestionably more important: ensuring a minimum standard of living for its citizenry.

It appears that this objective is attainable through a reform of European institutions whereby they would focus on the needs of people rather than the needs of the economy. It is thus necessary for Europe to decide once and for all to be what everyone wants it to be: a union of destinies rather than a union of markets. This entails preparing to integrate our several national realities and renounce any form of egocentric behavior that may induce us to forget that the EU has given its people the longest period of peace and prosperity ever experienced in Europe's long history.

The data throughout this chapter demonstrate that the effects of the recession have been especially severe over the course of recent years. The causes—as always in such cases—can be traced back to the country's recent past, to the period preceding the onset of the crisis in 2008. First, the illusion of an economy undergoing continuous growth paralyzed any political initiative aimed at rebuilding Spain's productive model on a more secure footing. The increase in GDP, which extended from the mid-1990s until 2007, led people to think that the Spanish economic wheel would continue to turn, come what may. These years of economic prosperity did not mean stable jobs, however. It is evident that the recurring economic crises in Spain have always had repercussions within the labor market, revealing not only a number of structural weaknesses in the Spanish model, but also another harsh reality: the fact that constitutional provision may not be the solution to a social problem unless effective legislation is enacted and implemented. Thus, while it is true that the constitution is committed to achieving full employment, this text, fundamental and very important as it is, has not been able to prevent rising unemployment levels, stemming from the slowing economy.

Although some initiatives, such as those created by the Andalusian government, may seek to address the plight of unemployed young people, these remedies seem to be both partial and temporary. It is clear that these initiatives result from a short-range political strategy, taking remedial measures to overcome a period of stagnation in the labor market, in the hopes that a revival in the economy will lead to a more permanent increase in employment rates. In order to achieve results that have lasting effects on employment levels, however, it is necessary to analyze and confront the root causes of unemployment, addressing on the one hand the educational system and, on the other, the performance of the labor market.

When looking at the education system it is crucial to remember that the right to education is constitutionally important because of the influence it has on the destiny of every social group. It is therefore unacceptable that an important region of Spain such as Andalusia continues to report record high levels of high-school dropouts. Educational efforts should thus focus on supporting at-risk students by promoting early-warning mechanisms and introducing mandatory participation in intensive and individualized classes. Additionally, there could be incentives to persuade students to remain in the education system, explaining and conveying both to young people and their families that investing in education is investing in access to employment. There should be flexible curricula that allow for the return of individuals who have dropped out of school early. It would also be undoubtedly beneficial to introduce mechanisms to facilitate direct transition from education to the labor market, bypassing unemployment altogether. These could include training programs carried out in collaboration with companies seeking employees. As far as the labor market is concerned, it is preferable that the training of unskilled workers be oriented to the real needs of the market. In this regard, it would be interesting to implement solutions that other European countries have already adopted, for example, Germany's dual vocational training systems. It is also necessary to invest in retraining courses.

The synergy between businesses and workers' representative bodies is key to tackling unemployment. It is therefore essential that, in developing plans for future employment, both bodies are actively engaged, with the interests of both represented equally. From this perspective, it is worth noting that the amount of money invested in implementation is not as significant as the investment patterns themselves. Palliative solutions are strategies that in the long term have proven ineffective and even harmful.

The picture we have detailed throughout this chapter allows us to understand why phrases like "jobless generation," "lost generation," or "NEET generation" are used to describe the current situation of many Spanish and Andalusian young people. These expressions indicate identical social phenomena that many young Spaniards born after the 1990s experience in their lives: insecurity, labor fluctuation, social immobility, and, ultimately, a lack of hope for the future. It is essential to remember that, just by providing an opportunity for these generations that we consider "lost," we will be able to secure a better future for all members of Spanish and Andalusian society. Only through political and common commitment and real action can we ensure that the constitutional and statutory provisions described do not become empty promises.

References

Agudo Zamora, M.J. 2007. *Estado social y felicidad. La exigibilidad de los derechos sociales enel constitucionalismo actual* [*The Welfare State and Happiness. The Enforceability of Social Rights in Contemporary Constitutionalism*]. Madrid: Laberinto.

Andalucía Podemos. 2015. "Programa Andalucía Podemos" ["The Political Platform of Andalucía Podemos"]. Retrieved from https://es.scribd.com/document/262552103/Programa-Podemos-Andalucia-2015

Bank of Spain. 2018. "Main Macroeconomic Magnitudes." Retrieved from www.bde.es/webbde/es/estadis/infoest/series/ie0102.csv (accessed June 26, 2018).

Capsada Munsech, Q. 2014. "Educación y desempleo juvenil" ["Education and Youth Unemployment"]. *Información Comercial Española, ICE: Revista de economía*, 881: 51–66.

Ciudadanos. 2015. "Programa Elecciones Andalucía" ["The Political Platform of Ciudadanos Andalucía"]. Retrieved from https://documentop.com/programa-elecciones-andalucia_5a4f85181723dd4cb40f5cd5.html

Contreras Peláez, F.J. 1994. *Derechos sociales. Teoría e ideología* [*Social Rights. Theory and Ideology*]. Madrid: Tecnos.

EUROFOUND. 2012. "Fifth European Working Conditions Survey." Retrieved from www. Eurofound.europa.eu/es/publications/report/2012/working-conditions/fifth-european-working-conditions-survey-overview-report (accessed January 23, 2018).

EUROFOUND. 2016. "Exploring the Diversity of NEETs: Country Profiles." Retrieved from www.Eurofound.europa.eu/sites/default/files/ef1602en2.pdf (accessed January 23, 2018).

European Council. 2016. "8.1 Recommendation CM/Rec (2016) 7 of the Committee of Ministers to Member States on Young People's Access to Rights." Retrieved from https://search.coe.int/cm/Pages/result_details.aspx?ObjectId=09000016806a8f5b (accessed January 23, 2018).

European Union (EU). 2016. "News Release, Euroindicators, 98/2015—3 June 2015." Retrieved from http://ec.europa.eu/eurostat/documents/2995521/6862104/3-03062015-BP-EN.pdf/efc97561-fad1-4e10-b6c1-e1c80e2bb582 (accessed January 23, 2018).

EUROSTAT. 2016. "European Labour Force Survey." Retrieved from http://ec.europa.eu/eurostat/web/microdata/european-union-labour-force-survey (January 23, 2018).

EUROSTAT. 2017. "Early Leavers from Education and Training." Retrieved from http://ec.europa.eu/eurostat/statistics-explained/index.php/Early_leavers_from_education_and_training (accessed January 23, 2018).

Faci Lucía, F. 2011. "El abandono escolar prematuro en España" ["The Early School Dropout Crisis in Spain"]. *Avances en supervisión educativa: Revista de la Asociación de Inspectores de Educación de España.* Retrieved from https://avances.adide.org/index.php/ase/article/view/468/312

Fernández García, T., and S. Andrés Cabello. 2015. "Crisis y estado de bienestar: las políticas sociales en la encrucijada" ["Crisis and the Welfare State: Social Policies at the Crossroads"]. *Tendencias & Retos*, 20(1): 119–132.

García López, J.R. 2014. "El desempleo juvenil en España" ["Youth Unemployment in Spain"]. *Información Comercial Española, ICE: Revista de economía*, 881: 11–28.

Molina Navarrete, C. 2012. "Art. 26 Trabajo" ["Art. 26 Right to Work"]. In *Comentarios al Estatuto de Autonomía para Andalucía* [*Commentaries on the Statute of Autonomy for Andalusia*]. P. Villalón Cruz and M. Medina Guerrero eds Sevilla: Parlamento de Andalucía, pp. 434–447.

National Institute of Statistics. 2017. "Labour Force Survey." Retrieved from www.ine.es/jaxiT3/Tabla.htm?t=4247&L=0 (accessed January 26, 2018).

Peces-Barba Martínez, G. 1993. *Derecho y derechos fundamentales* [*Law and Fundamental Rights*]. Madrid: Centro de Estudios Constitucionales.

Rahona López, M. 2008. "Un análisis del desajuste educativo en el primer empleo de los jóvenes" ["An Analysis of the Educational-Level Polarization of the Initial Employment of Youth"]. *Principios: estudios de economía política*, 11: 45–70.

Ramón García, J. 2011. "Desempleo juvenil en España: causas y soluciones" ["Youth Unemployment in Spain: Causes and Solutions"]. *BBVA Research: documentos de trabajo*, 11(30): 1–25.

Serrano, L., A. Soler, and L. Hernández. 2013. *El abandono educativo temprano: análisis del caso español* [*The Early School Dropout Crisis: Analysis of the Spanish Situation*]. Valencia: Instituto Valenciano de Investigaciones Económicas.

13 Bad schools, no jobs, full jails

Mass incarceration and a monumental incentive failure

Marcellus Andrews

Preamble: in remembrance of Bob Prasch

This chapter summarizes some results from a long program of work by the author on the connections between racial conflict, economic inequality, and the use of prisons to control the consequences of radically unequal opportunities by race in the United States. Bob Prasch once observed:

> It may be that America's taste for using jails to control crime is what a free market society does when it can't face the fact that lightly regulated capitalism is a miserable social failure; it is easier to blame black people for being poor because they are inferior—criminals by nature and nurture don't ya know—and then lock them up than admitting that the system is a disaster. Perhaps this is all just one giant mistake that our "leaders" and their public cannot face up to! Pride goeth before the Fall and all!
>
> (Prasch, personal conversation, early spring 2012)

I scoffed at Bob's off-hand remark—I am old, black, and scared for all the young black people in my life, especially my sons. But economic theory strongly suggests that Prasch may have been right—perhaps this is all one massive accident—in the parlance of economics, a vast incentive failure. I had not tried to apply the relentless logic of game theory and some well-known basic economics to the matter until Bob pushed me with his deep yet gently stated wisdom. The next few pages sketch my answer to Bob's quip—my complete answer is fully developed in a series of agent-based computational models that allow a full systemic analysis of the dynamics of racial fighting and inequality in an erstwhile market democracy. That model is best explored in a different venue. My goal in these pages is to show how ordinary economic logic, of the sort first learned in Econ 101, can lead to the terrible outcome of mass incarceration.

Bad jobs, no jobs, full jails, and the failure of self-interest

One of the good things about economists is that we are materialists—no matter our "politics." We search for the source of persistent social problems—in this case the troika of educational failure, youth unemployment, and mass incarceration—in the self-destructive choices of individuals, businesses, and governments. In this chapter I extend that search to explore a few elements of game theory and show how the US school-to-prison pipeline could be due to a series of linked incentive failures rooted in a complex mix of race- and class-based choices about where to live in a nation that relies on local control of education and policing. Hard-edged economics suggests that racial and class inequality in schooling contributes to economic inequality across color lines, one result of which is persistent differences in youth unemployment among racial groups that in turn drive race-based differences in criminal activity. It follows, then, that the decentralized nature of public goods provision in the United States, particularly in matters of schools and punishment, combined with short time horizons and other perverse incentives of electoral competition, encourage governments to favor punishment over education as a primary means of crime control. In a sense, mass incarceration is an example of how individual self-interest among tenants, landlords, homeowners, politicians, police officers, poorly schooled young people, and customers in illegal markets link up to create a monumental collective failure that resists reform.

Many Americans, especially black Americans, look at this state of affairs and conclude, reasonably, that America is a "racist" society in the strict sense that harmful social policies are pursued or at least tolerated because the victims of these policies are disproportionately black. But racism, though important, matters less than the persistent system of bad choices made by ordinary citizens as buyers and sellers in housing markets, as buyers in markets for illegal drugs, as voters, and the choices made as well by political authorities to favor punishment over genuine equal educational opportunity as a way to control the social costs of structural inequality. Even the most extreme but rational racist knows that managing school failure and unemployment through state violence is an expensive and ultimately fruitless proposition. Recent attempts to move America from its reliance on punishment, as made by conservative and right libertarian thinkers (well known for their indifference to racial equality), signal a recognition that there are strong limits to the use of brute force to control the social costs of inequality. Informed observers know that locking up poorly schooled young people to control crime makes crime worse in the future for reasons I examine below.[1] Yet the United States, as a nation, is unable to move away from the toxic mix of structural inequality and race-based mass incarceration because all of the major players in the game are trapped by their own self-interest, much like Prisoner's Dilemma leads to the worst possible outcome when everyone acts on their dominant strategy.

Stylized facts about mass incarceration

We can fix ideas by considering some of the empirical connections between schools, jobs, and mass incarceration. Bruce Western (2007) and Western and Petit (2010) have pursued some of the most extensive studies of the connections between economic inequality and mass incarceration in the United States, in the process painting a nightmarish portrait of social despair at the bottom of the system that has a direct bearing on the problem of youth unemployment. One particularly comprehensive summary of these studies provides a compact yet devastating portrait of the connections between poverty, inequality, race, and punishment that highlights the vicious cycle between youth unemployment and prison (Western and Pettit 2010). As Western and Pettit note: "The social inequality produced by mass incarceration is sizable and enduring for three main reasons: it is invisible, it is cumulative, and it is intergenerational" (2010: 8). The damage done by mass incarceration, they write, "is invisible because it is inflicted on populations that are usually outside the official accounts of economic well-being": primarily populations of young men outside the labor force (Western and Pettit 2010: 8). The damage is cumulative because the involved populations tend to be drawn from the bottom of the economic and wealth distribution system as well as from outcast populations, especially black Americans. Finally, the damage is intergenerational because mass incarceration generates important negative externalities that affect families and especially children (Western and Pettit 2010: 8). Table 1 of Western and Pettit's paper (reproduced as Table 13.1 here) shows what the authors call the cumulative risk of being imprisoned between age 30 and 34 for men born between 1945 and 1949, and between 1975 and 1979, by race and educational attainment. This table, derived from the National Longitudinal Survey of Youth, sponsored by the Bureau of Labor Statistics, US Department of Labor (2006, as cited in Western and Pettit 2010), is one of the most startling pieces of statistical information in the matter of mass incarceration that has a direct bearing on youth unemployment. The cumulative nature of these probabilities means that the figures in the table are estimates of the lifetime chances that men in different racial and educational groups will go to prison by the time they reach the 30–34 age range.

Table 13.1 tells us a number of disturbing things about the impact of the American experiment in mass incarceration on the life prospects of men over

Table 13.1 Probability of incarceration by age 30–34 by race and educational attainment

	Overall	High-School Dropouts	High School/GED	College
1945–1949 Cohort				
White	1.4	3.8	1.5	0.4
Black	10.4	14.7	11.0	5.3
Latino	2.8	4.1	2.9	1.1
1975–1979 Cohort				
White	5.4	28.0	62.0	1.2
Black	26.8	68.0	21.4	6.6
Latino	12.2	19.8	9.2	3.4

Source: Western and Pettit (2010: 11).

their life cycle. First, men from all racial groups had a much higher likelihood of incarceration in the 1970s as compared to the mid- to late 1940s, with black men being particularly likely to spend a long time in prison as a consequence of offending behavior. These estimates are consistent with the increasing importance of formal schooling in a technology-driven society because educational failure has been far more likely to lead to economic and social exclusion since 1979. As the American experiment with mass incarceration became especially pronounced in the period from 1980 through the mid-2000s, with aggregate incarceration rates for white and black men rising from 168 and 1,111 per 100,000 in 1980 to 468 and 2,805 by 2013, respectively, according to the Bureau of Justice Statistics, the probabilities listed in Table 13.1 are likely to *underestimate* the probability of punishment in recent years (Correctional Populations in the United States 2015). While the 1980 estimates for mass incarceration by the Bureau of Prisons did not include data on Latinos as a separate category, the report for 2013 tells us that the incarceration rate for this population is 1,134 per 100,000. We can nonetheless use these data to make good guesses about the impact of punishment policy combined with youth unemployment on the economic wellbeing of young men. Second, the probability of punishment drops with higher degrees of educational attainment, although the racial disparities signaled in Table 13.1 are stunning, particularly for black men without benefit of a high-school diploma. All other things being equal, a black male high-school dropout is more than 2.25 times as likely to go to prison as a white dropout and more than 2.5 times as likely to go to jail as a Latino dropout.

We can combine the probabilities estimated by Western and Pettit (2010) with data on educational attainment and youth unemployment across color lines to get a rough sense of how youth joblessness and punishment can become a vicious cycle for men at the bottom of the system. There is a substantial amount of evidence that young male high-school dropouts have much higher crime rates than other populations and that, on average, prisoners tend to be among the least well-educated members of their generation. This latter point was confirmed by a study in 2003 showing that more than 41% of all inmates in the United States at that time had not earned a high-school diploma or general equivalency diploma (GED) (Harlow 2003). There is also overwhelming evidence that young black and Latino men exhibit much higher dropout rates than white men, particularly in communities with high degrees of concentrated poverty and high rates of educational failure. For instance, data from the National Center for Education Statistics (NCES), specifically the Digest of Education Statistics (DES), show that while the national high-school dropout rate in the United States for all young men between the ages of 16 and 24 in 2014 had fallen to a record low of 7.3%, the rate for white, black, and Latino men was 4.8%, 8.1%, and 13.9%, respectively (NCES/DES 2015: table 219.70). High dropout rates in turn lead to high unemployment rates among dropouts, thereby contributing to high crime and incarceration rates among the young and poor. Table 13.2 shows the unemployment rates for men in the 20–24 age group by educational attainment in 2014 from the Current Population Survey by way of the NCES/DES (2015). The table

Table 13.2 Unemployment rate by highest degree attained, 2014

Highest Degree Attained	Unemployment Rate
No High School	25.3%
High School	18.9%
Some College	12.2%
Baccalaureate	6.7%

Source: NCES/DES (2015).

confirms something that everyone knows about modern economic life: unemployment rates are inversely related to degrees of educational achievement, especially among the young. The unemployment rate in the 20–24 age range runs from 25.3% among those young people without a high-school diploma to a relatively small 6.7% for college graduates—although this incidence of unemployment is still a shock to those who might have thought of a college degree as insurance against joblessness. These facts, along with Western and Pettit's (2010) foregoing estimates of the lifetime probability of imprisonment for men reaching the 30–34 age range, tell us that young male high-school dropouts, especially black male dropouts, are quite likely to become prisoners, which greatly diminishes their chances of high wage employment, or any employment, once they leave confinement. Western (2007), in his book *Punishment and Inequality in America*, provides strong evidence that former prisoners will spend most of their working lives shuttling between the low-wage labor market, the illegal sector, and prison because employers are reluctant to hire them. Western (2007: 131–168) also provides evidence that mass incarceration undermines family stability by reducing the employment prospects of fathers and husbands, thereby increasing the proportion of families with children headed by women, and in the process putting severe economic and emotional stress on poor and working-class mothers.

The simple economics of the school to prison pipeline

The idea that a social nightmare as nasty as mass incarceration could be a collective action problem may seem insulting to millions of black Americans, including me, because it seems to downplay the role of racial animosity and overt racist politics in American life. But the stone-cold materialism of economics demands that we pay attention to the discipline's most basic lesson, namely that social calamities might be the end result of self-interested choices in the same way that Adam Smith taught us that collective prosperity can be the unintended consequence of personal greed. We are all too aware of how other collective failures, from climate change to the most recent global financial collapse and the self-destructive nature of free markets, can be due to incentive structures going wrong. Could the same be true for mass incarceration in a racially divided society? How can mass incarceration happen in a liberal society that officially eschews state-sponsored racial or caste hierarchies?

This is a difficult question that cannot be fully answered in a few paragraphs, but very basic game theory provides some useful clues. The following analysis extends the work of Kaushik Basu (2011), whose analysis of the brutal economics of identity offers some intriguing clues about how a regime of mass incarceration can be a logical if vicious outcome of ordinary economic life in a divided society.

The first part of an answer to the question posed by mass incarceration is recognition that class divisions in society mean that the well-off not only garner most of the benefits of economic cooperation but also suffer few of the costs as well. Well-schooled and rich citizens not only own a disproportionate share of a community's material and human assets, thereby earning the lion's share of income, but also have an outsized role in setting the political and policy agenda. The wealth of the well-off also shields them from the fallout from social inequality—the suffering that flows from economic and social inequality does not generally cast a shadow on the lives of the well-off because they do not live near the poor, at least in the sense of social distance. As anyone who has spent any time in New York, Los Angeles, Chicago, or any other major American city knows well, the poor, even the very poor, keep their distance from the well-off, even when beseeching people of means for spare change with which to buy another bit of intoxication.[2]

Second, working-class and poor people live in the midst of the suffering and chaos following on from social and economic inequality. Families and neighbors are all too frequently divided between those who struggle to accept their low place in society with dignity and grace, following the rule of laws they have no role in making or enforcing, while other family members and neighbors break the rules out of anger, desperation, frustration, or malice. The fight for well-being pits poor neighbors against each other, all too often in the form of alienated young offenders preying on each other and world-weary adults through all sorts of crimes. Life at the bottom can all too often become a cage match where the poor are left to fight among themselves in a constricted social space constructed in the interests of those who own society and establish policy.

The separation between the owner/rulers of society and the rest leads to a profound conflict of interest between the top and bottom of the system, with the top favoring security in the interest of keeping the machinery of economic cooperation going, while the bottom seeks not only safety from the costs of social inequality constructed at the behest of the owners of society but also opportunity for their own children. The top of society has only very limited interest in opportunity for the children of the bottom since, frankly, the rest of society is only of interest to the top to the extent that their cooperation is required for economic life to function well enough for the system's owners. If the bottom can push the top to meet its needs through elections, or through various forms of non-cooperation or even more direct fighting, then its needs will be met lest the entire system fall apart. But the bottom of society faces a daunting collective action problem: how to organize and sustain pressure on the top that forces them to consider the needs of the bottom in the context of the existing system? The needs of the bottom of society are many, beginning with the need for the protection of the law abiding from the lawless within the bottom. The security needs of the bottom of society from crime and social disorder are both real and far cheaper to meet than the fiscal and organizational requiremehts for genuine equal opportunity for development and achievement in society: it is far easier for the top of society to build prisons and mete out punishment for offenders than it is to create a society where the bottom has a real chance of leading good lives. And as the top of society is prone to the usual vanity about its intellectual and moral superiority *vis-à-vis* the bottom—as, of course, demon-strated by its wealth and social position—imprisoning the "bad" poor in the interests of protecting the "good" poor has a certain missionary and charitable aspect that preserves the hierarchy. The bottom of society is too poor and disorganized to mount a serious challenge to the top, one that forces real change in the structure of the system, and the top has the means and motive to use state power to offer protection for the "good" poor against the depredations of the "bad" poor without actually ever addressing the real problems of structural economic, racial, and social inequality.

We can represent the game between the top and bottom of society as a standard two-population matrix game as shown in Table 13.3. The row popula-tion comprise the owner/rulers of socicty while the column population repre-sents the rest of the population. Economic cooperation yields a level of output per worker (y) which is divided between the top of society that is $0 < m < 1\%$ of the population but earns $0 < b/m < 1\%$ of all income where $0 < m < b < 1$ is the fraction of all income that the dominant group earns. A fraction of total income $0 < \theta < 1$ is transferred from the top to the bottom of the system in order to provide public goods as well as some degree of income and opportunity redistribu-tion in accordance with the social contract. Note that θ_1 and $\theta_2 < \theta_1$ are the transfer percentages associated with egalitarian (Regime 1) and punitive policy (Regime 2) regimes, which are ultimately under the control of the top of society. The bottom of society earns $0 < 1{-}b < 1{-}m < 1\%$ of income as well as the transfer $\theta b/m$ from the top. The bottom can choose to resist the agenda of the top through protest, non-cooperation, or violence at a per-person cost of c_R or it can submit to the will of the top at a cost $c_S < c_R$. If the bottom chooses to resist the top it can impose a cost of $x > 0$ on the average member of the top, although we assume that the bottom would only resist the top if the latter opts for inequality over equality.

Table 13.3 The game of economic dominance and submission

	Resistance	Submission
Equality	$\frac{b(1-\theta_1)y}{m}$, $\left[\frac{1-b}{1-m} + \frac{\theta_1 b}{m(1-m)}\right]y - c_R$	$\frac{b(1-\theta_1)y}{m}$, $\left[\frac{1-b}{1-m} + \frac{\theta_1 b}{m(1-m)}\right]y - c_S$
Inequality	$\frac{b(1-\theta_2)y}{m} - x$, $\left[\frac{1-b}{1-m} + \frac{\theta_2 b}{m(1-m)}\right]y - c_R$	$\frac{b(1-\theta_2)y}{m}$, $\left[\frac{1-b}{1-m} + \frac{\theta_2 b}{m(1-m)}\right]y - c_S$

The top of society will opt for inequality and a primary tool of social control, mass incarceration, if the cost that the bottom can impose upon them (x) is less than the gains from imposing a lower cost yet violent regime on society—i.e., $x < \frac{b}{m}(\theta_1 - \theta_2)y$. But the top can only pursue inequality as its policy strategy so long as the bottom is unable to push the cost of resistance (c_R) below the hideous costs of submission (c_S), costs that include the costs of mass incarceration noted above. Whenever $c_S < c_R$, whether, as noted above, because the costs of organizing collective action are just too high, or the costs of resistance are too great due to state violence, or the owners of the system are clever enough to limit the costs of submission, the game between the top and bottom of society results in substantial inequality matched to submission. This brutal Nash equilibrium will be perfectly sustainable (or in the parlance of economics, a long-run equilibrium) so long as the bottom does not find ways to make resistance feasible.

If the top of society is especially stupid, it might act so brutally and without regard to the future that over time the costs of submission might rise to the point where $c_S > c_R$ and the dominant strategy for the bottom is to resist. Also, the bottom of society might have its own problems with stupidity, particularly forms of racism that encourage one subset of the wretched bottom dwellers to heap scorn and abuse on an especially despised sector of the bottom by succumbing to appeals about the "naturally" violent, bestial, shiftless, and sexually incontinent "inferior" races or immigrants. This racist calumny is a narcotic that lulls the "superior" race stuck at the bottom of society to sleep, in that way reducing the costs of subsisting at the bottom of the system by encouraging bottom-dwelling racists to support harsh punishment policies against people who, but for their color or religion or national origin, share the same low economic station of the racists.

Much of this analysis hinges on the idea that there is little solidarity between the top of society and everyone else in a racially and economically segregated society. A high degree of altruism would go a long way towards reducing the incentives towards mass incarceration: the well-off *could*, if they wanted to, include the role of racial and economic segregation, political failure, and the negative effects of punishment policies on poor communities in their assessment of policies and parties when voting and supporting candidates for office. But altruism is the product of both ethics and "experience" in the dual sense of a shared set of values, as well as a shared view of what we can rationally expect in the social world. A society that takes educational failure among the poor for granted, particularly among poor people of a different race or ethnicity, will not have reason to question public policies like mass incarceration, which would then seem to be the best public response to crime. Further, politicians facing dual demands for low crime rates and low taxes are only behaving rationally when they opt for punishment over better schools for the poor, not least because the benefits of more jails are obvious and immediate while those of opportunity for the poor are more distant and less certain.[3]

We should note that racism, in and of itself, is not required for a regime of inequality and resentful submission to persist. All that is necessary is that the cost of resistance is high enough compared to the costs of submission to make submission the dominant strategy of the bottom of society. But racial animosity at the bottom of the system aids the top in imposing its agenda on the rest of us by lowering the cost of submission for racist dupes—those who despite the evidence of their eyes and the vast body of scientific knowledge cling to the idea that they are better than the other people at the bottom who occupy the same low rungs on the ladder of wealth and power—while simultaneously raising the costs of resistance by making it ever harder to craft solidarity at the bottom of society.

Conclusion

Can mass incarceration end? Yes, but only if, as the game of economic domination and submission tells us, the bottom of society finds a way to reduce the cost of resistance below the costs of submission. The top of society is not so clever as to have designed the game of dominance and submission, but it is smart enough to use historic racial animosities and the power of wealth to solidify its position while poisoning relations among the various segments of a diverse and mutually hostile bottom population. The foregoing economic analysis suggests that mass incarceration is just one more tactic that the top of society uses to retain control of the system, just another lever of power. The bottom must decide whether and how to alter the game by changing the structure of payoffs, particularly by changing the costs of acquiescing to the status quo and fashioning new ways of imposing severe costs on the top for sponsoring social regimes that do so much unnecessary damage to so many for no ethically conscionable reason. The bottom does not have to accept its fate if, but only if, it recognizes the nature of the game.

Notes

1 Dagan and Teles (2016) explore the recent turn against mass incarceration among a small sector of American conservatives that can be traced to a growing realization that prisons are an ineffective way to control the costs of economic and social inequality as well as anti-statist attitudes much more consistent with a limited government perspective than the punitive approach pursued in prior decades.
2 A word about the terminology I use is in order because many readers, especially Americans, will find my use of the words "top of society" and "owner/rulers" offensive to their democratic, middle-class sensibilities when a moment's reflection on the nature of the American racial and class system will dispel a certain sentimentalism about "middle America." Few doubt that the political agenda in America is set by the richest families and the managerial elites of the nation, who shape politics and public discourse around the vital controversies of the day in light of their own interests. The vast bulk of American voters are so-called middle-class people—those in the middle 60% of the income distribution—whose interests must be attended to if politicians of the two major parties are to garner a majority of votes. The rich in America cannot get what they want without at least catering to the needs of the voters—while the rich have control of the agenda they do not have complete control of the electoral outcome. The study by Gillens and Page (2014), which claims to show, via statistical methods, that US elites get almost everything they want out of Congress and that middle-class Americans get almost nothing they want, has been challenged by a number of writers recently. But racial and class animosities in the United States drive an enduring wedge between the American middle class and the bottom of the system—especially the

brown and black bottom—in light of centuries of racial hatred written into the political imagination and social DNA of this society. The "top" of American society is therefore best understood in the sense of a dominant group seeking to control the political body of the state, managing, trapping, and if need be disposing of the bottom of the system in light of its own interests, including its demand that the United States remains a white nation under white control. The alliance between a thoroughly pleasant middle-class man or woman in Topeka, Kansas, and a rapacious Wall Street trader in their mutual disgust for and animus towards poor people, especially black people, creates common cause that recruits middle America to act as a top to the black bottom. All this suggests that there is a Sadean element to race and politics in American life that cannot be sanitized by analytical social science or wished away by appeals to the goodness of the American common man or woman. Any serious study of American history would confirm that sentimentalism in the face of the nastiness of American life is part of the way in which the top of society blinds itself to its own brutality.

3   The connections between racial and class sorting in housing markets, the choices of young poor people between legal and illegal work, political competition, and the incentive structures of policy formation, and the resulting complex of high unemployment rates for the young, dark, and poor are being fully developed through the use of agent-based modeling techniques rooted in complex adaptive systems models of market and social processes. See Miller and Scott (2007) as well as Miller (2016) for an introduction to why complex adaptive systems in conjunction with evolutionary game theory offer a powerful set of analytical tools with which to explore the dynamics of inequality and racial conflict emerging from ordinary economic and political self-interest.

References

Basu, Kaushik. 2011. *Beyond the Invisible Hand: Groundwork for a New Economics.* Princeton, NJ: Princeton University Press.
Correctional Populations in the United States. 2015. Bureau of Justice Statistics, Office of Justice Programs, US Department of Justice. Retrieved from www.bjs.gov/index.cfm?ty=pbdetail&iid=5870
Dagan, David, and Steven Teles. 2016. *Prison Break: Why Conservatives Turned against Mass Incarceration.* New York: Oxford University Press.
Gillens, Martin, and Benjamin Page. 2014. "Testing Theories of American Politics: Elites, Interest Groups, and Average Citizens." *Perspectives on Politics*, 12(3): 564–581.
Harlow, Caroline Wolf. 2003. "Special Report: Education and Correctional Populations." US Department of Justice, Office of Justice Programs. Retrieved from www.bjs.gov/content/pub/pdf/ecp.pdf
Miller, John. 2016. *A Crude Look at the Whole. The Science of Complex Systems in Business, Life and Society.* New York: Basic Books.
Miller, John, and Page Scott. 2007. *Complex Adaptive Systems: An Introduction to Computational Models of Social Sciences.* Princeton, NJ: Princeton University Press.
National Center for Education Statistics/Digest of Education Statistics (NCES/DES). 2015. Retrieved from https://nces.ed.gov/programs/digest/d15/tables/dt15_219.70.asp
Western, Bruce. 2007. *Punishment and Inequality in America.* New York: Russell Sage Foundation.
Western, Bruce, and Becky Pettit. 2010. "Incarceration and Social Inequality." *Daedalus*, 139(3): 8–19.

Index

References for tables are shown in **bold**, and those for figures are in *italics*.